RAINHILL MEN
RAILWAY PIONEERS

ANTHONY DAWSON

AMBERLEY

First published 2022

Amberley Publishing
The Hill, Stroud,
Gloucestershire, GL5 4EP

www.amberley-books.com

ISBN: 978 1 4456 9844 1 (print)
ISBN: 978 1 4456 9845 8 (ebook)

British Library Cataloguing in Publication Data.
A catalogue record for this book is available from the British Library.

Typeset in 10pt on 13pt Celeste.
by SJmagic DESIGN SERVICES, India.
Printed in the UK.

Contents

Introduction: The Rainhill Trials

The Rainhill Trials were one of the biggest public science demonstrations in the early nineteenth century. It is estimated that 10,000 flocked to witness the 'trial of locomotive carriages' on a level stretch of the still incomplete Liverpool & Manchester Railway. Not until the Great Exhibition of 1851 would any event capture the public imagination in the same way. The late eighteenth and early nineteenth century had seen a growth in 'public' science when people became more aware of science and the workings of the world around them. Popular authors such as Dr Dionysius Lardner were doing much to make science accessible. There had been a growth in literary and philosophical societies, whose members tended to be well educated, middle-class Nonconformists. Those of Liverpool, Manchester, and Newcastle Lit & Phils came to be dominated by freethinking Unitarians and the occasional Quaker. The growing Mechanics' Institute movement brought science and technology to the masses via public lectures for a nominal admission fee. Such institutes were entangled with local Nonconformity; for example, the Manchester Mechanics' Institute began in rooms at Cross Street Unitarian Chapel. The whole idea of what 'science' was was rapidly changing. It was only in the 1830s that science began to take on its modern meaning in the English language; until that time the word had denoted merely a skill at something. Hitherto, what we would now categorise as 'science' was termed 'natural philosophy' and was for those who systematically investigated the natural and physical world. Robert Stephenson, for example, studied natural philosophy, natural history and chemistry while at the University of Edinburgh.

For the few who could read, the Rainhill Trials were a literary event. It was covered in detail by the technical press such as the *Mechanics' Magazine*, and by metropolitan and provincial newspapers. It was not just British scientists and engineers who came to witness the spectacle: engineers from the United States of America, France, and Germany were in attendance, each reporting back enthusiastically in their home country. Rainhill was covered by the French technical press and was even covered internationally, including *Niles Weekly Register* in the USA.

Public reception of science and technology was mixed – they were both wonderful and terrifying. As the Industrial Revolution gathered pace, science and technology were all around, for good or bad. There were hundreds of smoking mill chimneys in Manchester and Wedgwood's kilns in Staffordshire. Society was undergoing a seismic shift with

The Rainhill Trials captured the public imagination like no other event until the Great Exhibition. This reconstruction, entitled 'Rocket Comes in First' from the *Globe* newspaper of 1888 tries to capture some of the excitement.

massive urbanisation, which disrupted the traditional links of land and lord. A burgeoning and educated urban middle class and 'merchant princes' came to challenge the existing squirearchy of the shires, leading to clamours for Parliamentary reform to enfranchise these new men and their new cities. Britain came the closest to outright revolution in the few years leading up to the Great Reform Act of 1832.

How the world was understood was also changing: Jean-Francois Champollion had translated the Rosetta Stone in 1824, revealing the world of ancient Egypt. Sir Charles Lyell published his *Principles of Geology* (1830–33), which overturned the belief in the biblical account of the creation of the world, unlocking 'deep time' and showing that the world was billions of years old. The first dinosaur footprints were found and described in Scotland in 1828 by Revd Henry Duncan, and Mary Anning found a fossil pterosaur at Lyme Regis in the same year. In France, Lamarck had promulgated theories of human evolution between 1802 and 1822, and the 'Lunar Men' – Erasmus Darwin, Joseph Priestley, Josiah Wedgwood, James Watt and others – had been pushing the boundaries of science, believing that through science and learning that they could change the world. Charles Babbage had demonstrated his 'difference engine', the first mechanical computer, in 1822. Mary Shelley, the daughter of the radical 'power couple' William Godwin and Mary Wollstonecraft (the pioneer of women's rights), had summed up many of the fears about science in her 1818 novel *Frankenstein*. In the novel, Victor Frankenstein personifies the Industrial Revolution and the new society it was creating: as a student, Frankenstein rejects the teaching and knowledge of an older generation and attempts to create life with cutting-edge science. To many readers this was shocking, setting up mankind as creator without any need for a divinity. While freethinkers such as the radical Unitarians welcomed these developments, they were not necessarily welcome – sometimes violently so. Old ways of thought and being were under attack; old certainties were being replaced by doubt. Even the old ways of travel were being challenged and superseded by the new railways. It was an exciting, challenging and, for many, daunting time to be alive.

The Rainhill Trials had been organised by the Board of the Liverpool & Manchester Railway in order to find the best motive power with which to work their line. They had approached the matter systematically. Members of the board were deputed to visit those railways that were locomotive worked and to make observations. They also visited those worked by horses and those worked by stationary engines and rope haulages. These visits, however, proved inconclusive, and the advice from the leading experts of the day did not help either. To come to some conclusion, in the autumn of 1828 and new year of 1829, James Walker of London and John Urpeth Rastrick of Stourbridge toured the country to make a technical study and report back to the board. Their report argued in favour of stationary engines and rope haulages. This infuriated George Stephenson and his allies – his son Robert, Henry Booth (the Secretary and Treasurer), Charles Tayleur, and Joseph Locke – who produced their own report repudiating many of the claims made by Walker and Rastrick, including the observation that locomotives could not climb hills. Stephenson even sought the counsel of Timothy Hackworth on the Stockton & Darlington Railway, which was worked by a combination of locomotives, horses and stationary engines. Hackworth urged him not to lose heart and expressed his own opinion that stationary engines would be completely unsuited to a great public line of railway.

Crucially Walker and Rastrick's report offered an optimistic suggestion for breaking this impasse:

> To enable you to take advantage of improvements which might be made [in steam locomotion]; with a view to encourage which, and to draw the attention of Engine makers to the subject, something in the way of a premium ... might be held to the person whose Engine should, upon experience, be found to answer the Best. The Rainhill [stationary] Engines would at the same time enable you to judge of the comparative advantages of the two systems.

Thus, in April 1829, the Board of the Liverpool & Manchester Railway announced a competition for the 'most improved' locomotive engine, 'which shall be a decided improvement on those now in use, as respects to the consumption of smoke, increased speed, adequate power, and moderate weight'.

As such, in order to ascertain the type of motive power to be used on the railway, the Rainhill Trials were held in October 1829. The Rainhill Trials were not only to help aid the choice of the best means of motive power available, but also to allow other engineers to showcase their work, to challenge what men like James Cropper feared would be a Stephenson monopoly on the L&M. The Rainhill Trials were as much a forum for the locomotive as it was for George and Robert Stephenson. Of the 'multifarious schemes' entered, four locomotives were finally selected: *Rocket* by George and Robert Stephenson and Henry Booth; *Novelty* by John Braithwaite and Captain John Ericsson; *Sans Pareil* by Timothy Hackworth; and *Perseverance* by Timothy Burstall. The winner, of course, was *Rocket*, which showed the triumph of the locomotive over rope haulages and stationary engines for working a timetabled public railway for the carriage of passengers and goods.

Above left: The stipulations for the Rainhill Trials were published in April 1829, setting out the requirements for entrants, including maximum weight, boiler pressure, cost and delivery date.

Above right: How the Rainhill Trials were to be organised was published on 6 October 1829. Each locomotive would make twenty trips on a measured course, the total distance representing a return journey from Liverpool to Manchester and back.

Right: The three leading contenders were *Rocket*, *Novelty* and *Sans Pareil*, depicted here by the *Mechanics' Magazine* to a common scale. The diminutive size of *Novelty* is readily apparent.

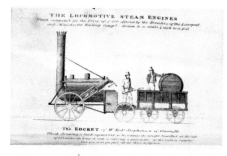

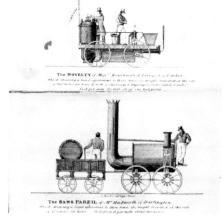

CHAPTER 1

The Judges

In order to oversee the Rainhill Trials, the board appointed three judges from among some of the most experienced engineers in the country: John Urpeth Rastrick, Nicholas Wood and John Kennedy. Often described as the 'lay' member of the panel, Kennedy had been trained as a millwright and machine maker, who, according to Edward Baines of Leeds, was 'well-known for his scientific attainments'.

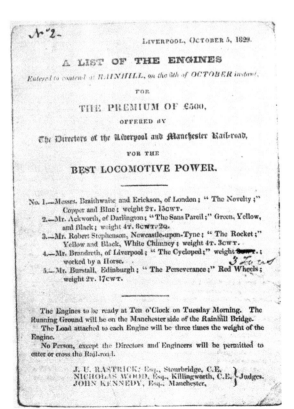

The three judges – Kennedy, Rastrick, and Wood – were three of the most experienced engineers of their day. Each of the three main contenders had their own striking 'racing colours': *Novelty* in blue, *Rocket* in yellow and *Sans Pareil* in green.

John Urpeth Rastrick

The Rastrick family originated in Yorkshire and were a family of engineers. Like many of the heroes of the Rainhill story, John Urpeth Rastrick (1780–1856) was a north countryman, born in Morpeth, Northumberland. He was the son of John Rastrick (1736–1828), 'an engineer and machinist of great ingenuity' employed in the 'construction of weirs, mills, and bridges'. J. U. Rastrick has been described as 'one of the most important engineers of his generation', although sadly his work has often been overshadowed by that of Thomas Telford or Robert Stephenson. At the age of fifteen he was apprenticed to his father and, after six years of pupillage, moved to Shropshire to work at the Ketley Ironworks. By 1811 he was a partner at the Bridgnorth Foundry, established by John Hazledine and his two brothers, Robert and Thomas. Rastrick took charge of mechanical engineering and was 'special charge of the iron foundry'. From *c.* 1802 the Hazledines had been doing work for Richard Trevithick, the pioneer of high-pressure steam and the steam locomotive. Hazledine and Rastrick built numerous high-pressure Trevithick stationary engines, of which No. 14 at the Science Museum, London, is a rare survivor. They also built Trevithick's locomotive *Catch Me Who Can* of 1808. John Hazledine died in 1810, and the partnership was dissolved in 1817. Rastrick entered into a new partnership in 1819 as managing partner of Foster & Rastrick of Stourbridge, which built steam boilers, steam engines, waterwheels and millwork. The most famous products of this partnership were three locomotives built for the American Delaware & Hudson Railroad in 1829, and a single locomotive *The Agenoria* for the Shutt End Colliery in Shropshire (1829).

As a civil engineer he designed and built the iron bridge over the River Wye at Chepstow, and was involved with various early railway schemes, including the Hay Railway for which

Hazledine & Rastrick of Stourbridge built Richard Trevithick's *Catch Me Who Can*, which was demonstrated in London in 1808. This iconic image by W. J. Welch forms the basis for a well-known print once believed to be by Thomas Rowlandson but later confirmed as a fake.

he designed a suspension bridge that was never built. His relationship with the Liverpool & Manchester Railway began in 1825 when he was consulted about the railways of the north-east coalfield and motive power. He also presented evidence in Parliament on behalf of the railway. Together with James Walker of London he toured the various railways of the North on a fact-finding mission on behalf of the L&M in winter 1828–29, reporting in March 1829 in favour of stationary engines and rope haulages, a report that directly inspired the Rainhill Trials. The L&M was one of several railways with which Rastrick was involved in the later 1820s, including the Stratford & Moreton and the Shutt End Colliery Railway (1827–29). He was also involved with the embryonic scheme to build a railway from Birmingham to Liverpool, which would eventually bear fruit as the Grand Junction Railway, which opened in 1837.

He was also engineer of the London–Brighton Railway (1840). Thus, at the time of the Rainhill Trials, Rastrick not only had experience of building a railway and railway locomotives, but was also well aware of their capabilities. He became a member of the Institute of Civil Engineers in 1827, a Fellow of the Royal Society and Member of the Society of the Arts. He retired in 1847 and died nine years later, aged seventy-seven.

The *Agenoria* was built in 1829, the same year as the Rainhill Trials, by Rastrick in Stourbridge. Her giraffe-like chimney is readily apparent.

Stourbridge Lion, named after the town of her 'birth', was the first steam locomotive to run in the United States of America.

Nicholas Wood

Wood (1795–1865) was the protégé of Sir Thomas Liddell, one of the 'Grand Allies' of north-eastern coal owners. He was trained at Killingworth Colliery as a colliery viewer – in modern parlance, a manager/technical manager – under Ralph Dodds (*c.* 1763–1821), with whom George Stephenson (see Chapter 2) had taken out a patent for a steam locomotive in 1815. Wood began his apprenticeship in 1811 aged sixteen, his talents having come to the attention of Sir Thomas, who offered him his apprenticeship. He would eventually succeed Dodds as viewer at Killingworth in 1815. Killingworth was a large, profitable, and wealthy colliery. Of 'commanding height' and 'portly form' Wood had a 'ruddy, good-humoured countenance, which bore no trace of the hard work he got through'.

The West Moor colliery and its associated railroad had been laid out in 1802–05, and it was at Killingworth that a series of factors came together to create the conditions leading to Killingworth being an early promoter and testing ground for the fledgling locomotive. The colliery was successful and needed something more powerful than horses to haul coal above and below ground. The Grand Allies were wealthy and the concern profitable and they could well afford to carry out experiments on steam locomotion with several like-minded individuals, including George Stephenson, working together to solve the same problem.

Above left: Nicholas Wood was a close friend and ally of George Stephenson and one of the most respected mining engineers of his day.

Above right: Wood was the 'viewer' here at Killingworth Colliery where George Stephenson would build his first locomotive, the *Blucher*, in 1814.

Stephenson took the younger Wood 'under his wing' at Killingworth and the two became firm friends and allies in the development, and advocacy, of the locomotive. Stephenson was an intuitive engineer, while Wood had the technical training and theoretical knowledge that Stephenson lacked; put together the duo were a veritable powerhouse of innovation. A young Robert Stephenson (1803–59) would be apprenticed to Wood in the trade of 'viewer'.

Wood would no doubt have been present for the first run of George's first locomotive, the *Blucher*, in July 1814 and his second, patent locomotive in 1815. The young Wood was also involved in Stephenson's development of the miner's safety lamp in 1816 and the ensuing controversy with Sir Humphrey Davy. Stephenson and Wood carried out experiments to understand rail vehicle dynamics in 1818–19, even constructing a dynamometer car. George recalled that Wood was one of the few men who did not abandon him after his rough handling and poor performance before a Parliamentary committee discussing the first Liverpool & Manchester Railway Bill in 1825. Wood was also of a mechanical turn of mind, and is credited with the invention and introduction of the slip-eccentric for working the valve gear of steam engines and locomotives. Wood also experimented with wheels and springs of early locomotives. He concluded that larger wheels were more efficient, lowering the rolling resistance of the locomotive and increasing its speed, and he adopted 4-foot-diameter wheels at Hetton Colliery in 1822. He then looked at the design of the wheel and introduced rolled, wrought-iron tyres to cast-iron railway wheels, which made them more durable. He also recognised the benefit of the tread of a wheel being coned, which helped lower the friction between the wheel and rail. Coning also helped keep the wheel on the rail and aided it pass round curves. Finally, he also introduced steel leaf springs to improve the ride of locomotives over brittle, cast-iron track, thereby improving the durability of both rails and wheels. Many of his experiments and locomotive developments were published in his *Treatise on Rail-roads and Interior Communication* of 1825, which brought him national attention. His

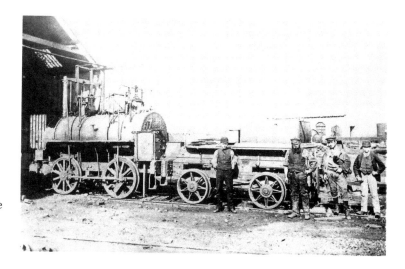

Many early locomotives had long lives. *Billy* is seen here at work at Hetton Colliery sometime in the 1860s.

Treatise, however, had to undergo a strict editorial policy before it could be published. Thus, by 1829 Wood had been a staunch supporter of George Stephenson and the locomotive for over a decade. He was one of the most experienced men on locomotive matters in the country, and therefore ideally suited to adjudicate at Rainhill. It is also thanks to Wood that *Rocket* had running-in trials at Killingworth in order to smooth out teething problems that the locomotive might have had, so that it was in the best condition possible at Rainhill.

Wood's 'authority as a mining engineer and geologist' was renowned across the north of England. 'His knowledge and experience caused his assistance to be sought' and he 'gradually acquired influence in the [coal] trade, possessed by few individuals, which continued undiminished until his death'. He was the first president of the North of England Institute of Mining Engineers, which honoured him posthumously with the Nicholas Wood Memorial Hall, in which stands a monumental statue of him. He was also a Fellow of the Royal Society. His four sons all made careers in the coal industry, with Sir Lindsay Wood (1834–1920) becoming chairman of the Hetton Collieries (of which Nicholas Wood had been a proprietor from 1844), being elevated to the peerage in 1897.

John Kennedy

Kennedy (1769–1855) was a Scot, born in Kirkcudbright in 1769, the son of a farmer. His father died when he was young so he and his four brothers were raised by his mother. At fourteen he was apprenticed to William Cannan, a machine maker of Chowbent, near Atherton in Lancashire. Cannan was a 'mechanic of some eminence', and it was thanks to his business partner James Smith, who was then installing cotton-spinning machinery in a Carlisle textile mill, that the young Kennedy moved to Lancashire. John's brother, James, also moved south and later commenced his own successful textiles business in Manchester. James McConnell, Kennedy's future business partner, also from Kirkcudbrightshire, was the nephew of Cannan and had been apprenticed to him, leaving his indentures in 1781 – three years before Kennedy was articled.

John Kennedy of Manchester is often considered to have been a 'lay' member of the judging panel, but this belies his considerable engineering experience as a millwright, carrying out his own development work on cotton spinning and weaving machinery.

McConnell and Kennedy entered their first partnership in 1791, and their second in 1795. Ultimately, the firm of McConnell and Kennedy of Ancoats, Manchester, would be one of the largest cotton spinners in Manchester – at its height employing 1,500 people. Both men were not just interested in spinning thread. They had both been apprenticed as millwrights and machine makers and thus had a mechanical turn of mind. The first spinning mule had been developed by Samuel Crompton in 1780, and Kennedy's firm became one of the leading builders of such machinery and spinning equipment. Kennedy himself carried out his own developmental work on the spinning mule, including the 'double speed' mule, which facilitated the spinning of very fine thread. McConnell & Kennedy were among the first in Manchester to use steam, employing a 45-hp Boulton & Watt condensing engine in 1803. In 1809 the firm began making its own 'town gas', and the mill was one of the first to be lit by gas.

Both men were Unitarians. Kennedy was a member of Cross Street Chapel where Revd William Gaskell (1805–84) was minister (1828–84) and he became friends with Mrs Elizabeth Gaskell, the novelist. McConnell attended the more theologically and politically radical Unitarian chapel on Moseley Street (the second avowedly Unitarian Church in Britain) and later donated a considerable sum to the replacement Gothic building by Sir Charles Barry on Upper Brooke Street. Kennedy was a member of a socially exclusive, industrial and entrepreneurial network of innovators and social reformers. But, because Unitarian belief was illegal until 1813, they were social outcasts. Unitarians had a level of wealth and influence in centres such as Birmingham, Manchester, Leeds and Liverpool that was disproportionate to their small number, among whom were such engineering luminaries as Peter Ewart (1767–1842), William Fairbairn (1789–1874), Eaton Hodgkinson (1789–1861), John Hawkshaw (1811–91), James Kitson (1807–85), and Richard Peacock (1808–89). James Watt (1736–1819) was also a Unitarian; his son James Jr was a member at Cross Street Chapel in Manchester. Josiah Wedgwood (1730–95) was also a Unitarian, and he and Watt were friends with Revd Dr Joseph Priestley, the Unitarian minister who discovered oxygen.

Kennedy was a member of the influential Manchester Literary & Philosophical Society where membership was not restricted to Unitarians, but it certainly helped. The society

McCONNEL & CO LTD
ANCOATS MILLS
MANCHESTER
1913

Above and right: Mconnell & Kennedy owned one of the largest cotton-spinning mills in Manchester. At its height they employed 1,500 people. The vast scale of the mill complex is shown in this early twentieth-century bird's-eye view. The Royal Mill is today a mix of apartments and commercial space in the chic New Islington development.

had been founded in 1799 and the first meetings were held at Cross Street Chapel. He was also involved in the Mechanics' Institute movement; the Mechanics' Institutes were envisioned to provide the 'mechanic and artisan' with an elementary education in the 'three Rs', as well as mechanics, chemistry, engineering and 'self-improvement' in an atmosphere free of sectarianism (political and religious). Among the papers Kennedy read to the Manchester Lit & Phil were on the cotton trade (1815), the Poor Laws (1819) and the impact of mechanisation on the working class (1826). He also wrote a memoir of Samuel Crompton, inventor of the spinning mule, in 1830.

Kennedy was among the first Manchester businessmen who promoted the Liverpool & Manchester Railway (1822) and was the only member of the provisional committee with any engineering or technical knowledge. He was part of the delegation that toured the various railways of the North East in 1824 and it was no doubt thanks to his Unitarian connections that many leading Manchester Unitarians subscribed to the new railway company, including Robert Hyde Greg (1795–1875) of Quarry Bank Mill, or William Potter of Manchester.

Kennedy retired from business in December 1826 to pursue his own mechanical interests and was well known in engineering and scientific circles; his share in the business was then valued at £85,000. In an obituary by his friend Sir William Fairbairn:

> Every discovery in mechanical science received his cordial support. He was a friend and admirer of Watt, and there were few distinguished scientific men with whom he was not acquainted, and on terms of friendly intercourse. Round his table at all times were to be found men who were noted for intellectual attainment ... Mr Kennedy never pursued business for the sake of money but for the love of improvements in his favourite mechanical pursuits. To these he devoted nearly the whole of his time ... He was fond of mechanical discussion.

These were excellent qualities to stand him in good stead as a judge at Rainhill. He died an extremely wealthy man at Ardwick Hall, aged eighty-six, in October 1855. His only son not having followed his father into the firm, the business was carried on by the McConnells, who invested in Welsh slate in the 1860s during the Lancashire Cotton Famine, purchasing a slate mine at Bryn Eglwys in Gwynedd in 1864, served by the TalyLlyn Railway, which opened in 1866.

Cross Street Chapel in the heart of Manchester was the nineteenth-century powerhouse of that booming industrial city. Sadly, this building, which dated from the late seventeenth century, was destroyed in the Blitz. (Cross Street Chapel)

CHAPTER 2

George Stephenson

George Stephenson (1781–1848) was the second of six children born to Robert and Mabel Stephenson. Old Robert was the 'fireman' of one of the pumping engines at Wylam Colliery and, as his biographer Samuel Smiles (1812–1904) records, the Stephensons and their large family lived in 'straightened circumstances', 'the father's wages being barely sufficient ... for the sustenance of the household.' Much of George's early life is shrouded in myth: George recalled that his first job, at the age of eight, was as a plough boy, and then working the horse gin at Black Callerton Colliery. Later George and his elder brother James (1779–?) were employed as Robert's assistant firemen at Newburn Colliery and this set him in good stead to be 'plugman' on the pumping engine at Water Row, responsible for the working of the engine and especially the pump gear. His father was the fireman on the same engine. A younger brother, Robert (1778–1837), later became involved with early railways, becoming 'engineman' on the Kenton & Coxlodge Railway (which used Blenkinsop/Murray-type rack locomotives) c. 1815, and subsequently worked at Hetton Colliery before becoming involved with William James (1771–1837) and the engineer Robert Wilson of Gateshead (1781–?) on the Stratford & Moreton Railway. Finally, he ended his career on the Bolton & Leigh Railway.

George Stephenson, the 'father of the railways' – one of the most familiar names and faces of the early railways.

Left and below: George Stephenson's birthplace in Wylam, a small two-storey cottage where Robert and Mabel Stephenson raised their six children. It is now owned by the National Trust. (Paul Dawson)

The young George had a natural flair for mechanics: according to Samuel Smiles, George often took an engine to pieces and reassembled it in his spare time and the engine 'became a sort of pet with him'. Remembering his early life at a speech made at Leeds in 1839, he recalled he saw his first locomotive 'about 30 years ago', but it's not clear what machine this may have been. Although the Wylam Waggonway literally ran past the front door of his childhood home, locomotives didn't begin working there, at least experimentally, until 1813 after George had left the village of his birth. It's possible, however, that this was the Trevithick-type locomotive built at Gateshead in 1805 by John Whinfield, perhaps speculatively for the Wylam Waggonway. Indeed, Richard Trevithick (1771–1833) recalled bouncing a young Robert Stephenson on his knee at the family home in Killingworth, *c.* 1805.

He married in 1802 and moved to work at Willington Quay as 'brakesman'. Around 1804 he moved to Killingworth and *c.* 1810 he came to the attention of the colliery viewer Ralph Dodds (*c.* 1763–1821) by modifying a pumping engine. The engine 'would not draw' but by raising the water tank for the condenser 10 feet, increasing the diameter of the injection cock, and raising the boiler pressure the improved engine was able to pump dry the flooded mine workings, and a grateful Dodds gave Stephenson a gift of £10. This was to be the beginning of a fruitful working relationship.

The Wylam Waggonway of 1748 ran past the Stephensons' front door. Part of the route became a section of the North Wylam Branch and this late nineteenth-century photo montage shows a North Eastern Railway express locomotive thundering past. It's no wonder George and his siblings were interested in railways.

As a young man George was known for his feats of strength. He enjoyed wrestling and throwing the hammer – in this case a literal hammer over the engine house roof!

Why Killingworth?

It was at the Kilingworth Collieries of the Grand Allies – Sir Thomas Liddell (later Lord Ravensworth), the Earl of Strathmore, and Mr Stuart Wortley (later Lord Wharncliffe) – that Stephenson made his mark as a skilled mechanic and in 1812 was appointed engine-wright on a salary of £100 a year and given the use of a pony to inspect the other pits owned by the Grand Allies. It was at Killingworth that Stephenson built up a team of engineers around him, some of whom would stay with him for the remainder of his career. As Robert Hartley has described, George was surrounded by other 'brilliant minds working on related problems' and, crucially, had the support of his employers, and thus their financing, for his early locomotive projects.

George's first foray into locomotive building came in 1814. George had perhaps been inspired by reports of a paper presented by the Unitarian Minister Revd William Turner (1761–1859) about the Blenkinsop and Murray locomotives at the Middleton Railway during 1812. Stephenson's first locomotive was probably named *Blucher,* supposedly after the Prussian field marshal Prince Gebhard Leberecht von Blücher (1742–1819), but 'blucher' is also a Geordie dialect word for a large, heavy object or a northern word for a large pair of boots. The *Blucher* was a small 0-4-0 locomotive that borrowed many features from the work of Blenkinsop and Murray. It had a single, large-diameter flue through the

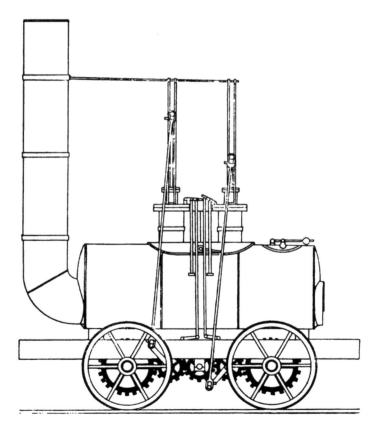

George's first locomotive was the *Blucher,* which was 'first tried' on 25 July 1814. It borrowed many key design features from Blenkinsop and Murray at Leeds.

boiler and vertical cylinders on the boiler centre line, part-immersed in the boiler in order to keep them warm. The piston rods drove upwards, guided by a cross head and slides. Final drive to the wheels was through a gear train.

It was first tried on 25 July 1814, a month or so before William Hedley's (1779–1843) locomotive at Wylam. It was the first successful adhesion-worked locomotive after Richard Trevithick's pioneering attempts. At first the steam vented directly to the atmosphere, but *Blucher* was often short of steam, so in order to increase steam production Stephenson, based on the experiences of Trevithick a decade earlier at Penydarren, routed the exhaust steam into the chimney through a pipe with an upturned end. This 'at once doubled the power of the engine, enabling her to go six miles an hour and maintain her steam'. Like Trevithick, George used a single, probably slightly coned blast pipe located in the centre of the chimney. But while this primitive form of blast pipe helped steam generation, it also presented a problem: noise. Nicholas Wood described a 'very disagreeable noise' as the steam was ejected into the chimney at each beat of the engine, which scared horses and upset neighbouring landowners. So, Stephenson had to soften his blast so that it would still draw the fire and help raise steam but also avoid the nuisance of noise. The other problem was that in using a single, large-diameter boiler flue, a strong blast would have ejected most of the coal, unburned, out of the chimney.

This hand-turned lathe on display at the Beamish Open Air Musuem in the North East was once owned and used by George Stephenson.

The Geordie Lamp

Locomotive development was put to one side while George tackled the problem of firedamp (methane gas) found in coal mines. There had been two serious explosions at the West Moor pit, in 1806 and 1809, resulting in twenty-one fatalities. One solution to the problem of using naked flames in an environment where pockets of undetectable methane were likely to form was presented by Dr Clanney of Sunderland in 1813, but it was a very unwieldy apparatus. George then turned his mind to solving the problem and, through practical experiment (including holding a candle up to a 'blower' leaking methane in one of the pits), he concluded that if the flow of air to the flame could be increased it would not ignite the flammable gas. His friend Nicholas Wood helped him sketch out the initial idea and the prototype was made by a Newcastle tinsmith. It had a long, glass chimney and had a perforated outer casing. It was tested in October 1815 by being taken down the mine to an area where there was methane gas hissing from a fissure and, although it blew out once, the lamp did not ignite the gas when relit. Further rapid development took place; the second was made and tested on 4 November and the third was ordered sixteen days later. But while this had been going on up north, in London Sir Humphrey Davy also presented his own solution to the problem: the 'Davy Lamp'. Davy and the London 'experts' regarded Stephenson's essay as 'the clumsy efforts of an uneducated man' and could not believe that Stephenson had come to a similar solution to the same problem as Davy. Influential colliery owners around Newcastle begged to differ, however, and presented Stephenson with an engraved silver tankard and a handsome purse of £1,000 raised through subscription. Davy saw this as a personal affront and sent letters to each of Stephenson's prominent supporters urging them to withdraw their claims, which they refused to do. This would start George's long-standing distrust of London 'experts' and establishment men, and, sadly, 'left an indelible stain' upon Davy's reputation.

In developing his miner's safety lamp George first took naked candles, then experimental models of his lamp into mines to observe how the flame reacted. As Tom Rolt states, George was incredibly brave, but also knew that no 'controlled experiment could ever convince the men on the job'.

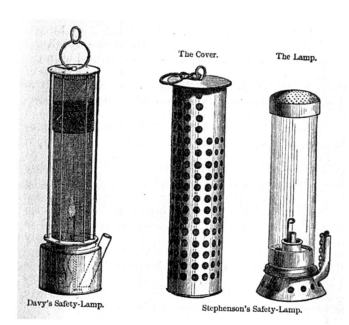

George's safety lamp, which used a perforated metal cover to prevent explosions, was developed on a trial-and-error basis contemporaneously to Sir Humphrey Davy's experiments, using a fine mesh.

The Cover.   The Lamp.

Davy's Safety-Lamp.   Stephenson's Safety-Lamp.

Patent Locomotives

To return to the locomotive story, within twelve months of building the *Blucher,* Stephenson had built a second locomotive and, together with Ralph Dodds, had taken out a joint patent for a locomotive in February 1815. The cost of the patent (£158 12s 4d) was paid for by their employers, the Grand Allies, who must have been suitably impressed with their work. Just a week later (6 March 1815) Stephenson's second locomotive had started work. It was larger than *Blucher* and, crucially, the geared drive was dispensed with; instead, drive was via long connecting rods from piston rod to wheel. To ensure cylinder synchronisation the wheels were coupled with chains, although Stephenson had originally proposed the use of crank axles and inside coupling rods but the technology of the day meant that such axles were unreliable and prone to fracture. This patent, however, may have been rendered null and void as a locomotive 'of the same construction' had been run on the Newbottle Waggonway before the patent had been published. The similarity in construction was probably the use of chains to couple the wheels, which had first been proposed by William Chapman (1749–1832) of Newcastle in 1814.

Stephenson took out a second patent in 1816, this time jointly with the Unitarian ironmaster William Losh (1770–1861), with whom he probably became acquainted via Revd William Turner (see **pp. 30–31**), who had taken the young Robert Stephenson (1803–59) under his wing (see Chapter 3). The Losh family were members of Turner's congregation at Hanover Square. William Losh had studied metallurgy in Sweden and chemistry in Paris, and established a chemical works in 1802 and an iron foundry and engineering works (Losh, Wilson & Bell) in 1809. Indeed, in 1816 George had been given leave by the Grand Allies to work two days a week for Losh, Wilson & Bell. George Stephenson became friends with the Loshes and would often spend a pleasant Sunday afternoon with them at their elegant town

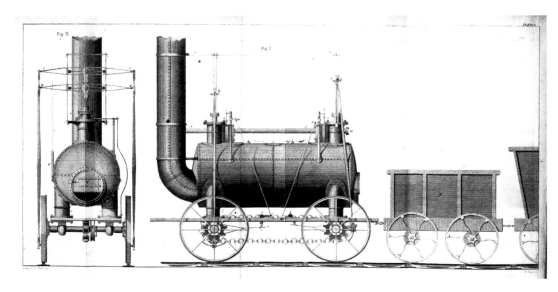

Stephenson & Losh's patent locomotive of 1816. Key design features included chains to couple the wheels and the use of 'steam springs' as a rudimentary form of suspension.

house. The Loshes were wealthy and influential: the brother, James Losh (1763–1833), was Recorder of Newcastle. The Loshes were radical in their politics and involved in the abolition of the slave trade and push for social reform. Sara Losh was a pioneering feminist, antiquarian and architect. Friendship with the Loshes and the well-respected mining engineer and locomotive pioneer John Buddle (1773–1843) – who was also a member of Hanover Square Chapel – certainly made useful allies. Indeed, as Andy Guy has noted, it appears Buddle and his partner Chapman were involved with Stephenson's first patent, as requests from Russia for drawings of the locomotive were passed to Buddle for permission. Chapman and Buddle also essayed several locomotives, including the *Steam Elephant* of 1815.

Stephenson and Losh's patent covered not only the locomotive but the rails it ran on too – George had a system-wide view of problems and saw that if the locomotive was to develop from a lumbering curiosity, then the track it ran on had to be developed as well. This was also true of the wheels, with George referring to the wheel and rail as being like 'man and wife'. The patent cast-iron rails used half-lap joints to make the ride smoother, and the patent also covered the unsatisfactory 'steam spring'. Before sufficiently strong steel leaf springs had been developed (they were first applied to locomotives by Nicholas Wood in 1827), Stephenson used a steam cylinder and piston to provide a rudimentary form of suspension. Sadly, they simply didn't work on the boiler pressure then available but were a brave attempt to reduce rail breakage and damage to locomotives and especially wheels from rough riding.

In his evidence before Parliament for the first Liverpool & Manchester Railway Bill in 1825, George states he had built some sixteen locomotives since 1813. He was thus the most experienced locomotive engineer in the country – perhaps world – at the time of the Rainhill Trials in October 1829. George built five locomotives for the Killingworth Railway between 1816 and 1821: two in 1816, two more in 1818 and the fifth in 1821. Another five, slightly smaller, engines were built for Hetton Colliery in 1822–23. But he did not just

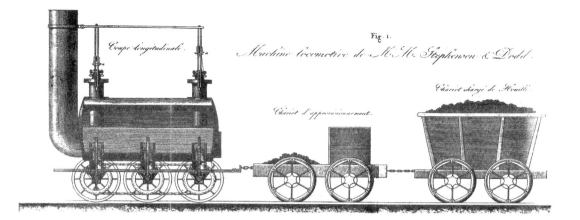

Above and right: Cross-section and elevation of the Stephenson & Losh patent locomotive showing the large-diameter single boiler flue, relatively long, strong cylinders and the working of the unsuccessful 'steam springs'.

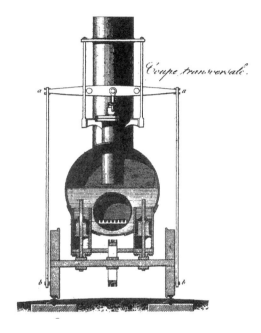

Stephenson and Losh's 1816 patent also included iron railway wheels as well as a form of cast-iron rail with a half-lap joint, which made the ride far smoother, leading to higher speeds and fewer broken rails or wheels.

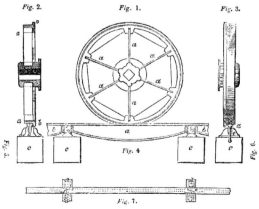

25

build locomotives for collieries in North East England. In 1816, George built a locomotive for the Kilmarnock & Troon Railway and in 1819 a locomotive for the Llansamlet Colliery in South Wales. The Kilmarnock engine was the first in Scotland. It was a six-wheeler with chain-coupled wooden wheels, and later found its way to the Paisley & Renfrew Railway. It was sold for scrap in 1848. The 1819 Llansamlet engine did not remain in use for long, as in 1824 proposals were being made for it to be used to pump water in a colliery. The last two locomotives George had a hand in the design of were the first two delivered to the Stockton & Darlington Railway and built by Robert Stephenson & Co. (see **pp. 32–36**) in 1825. These were the first locomotives to use outside coupling rods, these supposedly being an innovation of James Kennedy (see p. 90). George was also involved in the design of *Experiment* (1827), which was an attempt to return to the use of horizontal cylinders in order to reduce an effect known as 'hammer blow', which was damaging to the track.

Due to George's association with Hanover Square Chapel, it should not be surprising that all of his pupils were Nonconformists: William Allcard (1808–61) and John Dixon (1795–1865) were both Quakers; Henry Clarkson (1801–96) and Frederick Swanwick (1810–85, later his personal assistant) were Unitarians, with Swanwick originally having been intended for the Unitarian

STEPHENSON'S LOCOMOTIVE, WORKED ON THE HETTON COLLIERY.

George built five of his patent locomotives for the Hetton Colliery in 1822–23. These were smaller and more compact than his larger Killingworth locomotives.

Billy, now preserved at the Stephenson Railway Museum, is considered to be the second-oldest railway locomotive in existence, having originally been built *c.* 1816 at the West Moor Workshops of the Killingworth Colliery.

Ministry; and Joseph Locke, a Roman Catholic. Catholics suffered from similar discrimination as Protestant non-Anglicans at this time. Nor should it be surprising that Robert Stephenson took Charles Tayleur, from a good Shropshire Unitarian family, as his business partner. Simply put, no one respectable would apprentice their son to, or do business with, a Unitarian.

Self-made Man?

Samuel Smiles (1812–1904) held up George Stephenson as the epitome of the self-made man. While George was certainly a skilled and intuitive mechanic, it was thanks to his friend Nicholas Wood and the Pease family of Darlington that he came to national prominence. Although Smiles suggests George was working almost single-handedly on developing the locomotive, George notes in a speech given in York in 1839 that 'I had fortunately very able assistants' working with him. As an older man, he liked to surround himself with 'bright young things' like John Dixon or Joseph Locke:

A great number of talented young men, were brought up under me, the sons of the poor and rich have been in my service, and wherever talent had showed itself, whether the individual possessing that talent was poor or rich, I have used my effort to promote the interests of that person.

George also had an anti-establishment streak, claiming at the end of his life how he had 'dined with princes, and peers, and commoners – with persons of all classes, from the highest to the humblest' and concluded that 'if we were all stripped naked, there's not much difference between any of us'.

While he is perhaps the most famous locomotive pioneer, George was not working in isolation, with William Chapman and John Buddle, or William Hedley, essaying several locomotives of their own. Wood's seminal 1825 *Treatise on Rail-Roads* did much to publicise the work of Stephenson. It isn't an objective account, however, as it ignores the work of other key players (e.g. Chapman and Buddle) in the locomotive story; omits the explosion of William Brunton's *Mechanical Traveller* of 1815, perhaps because it would be prejudicial to the popularisation of the locomotive; and carefully omits the probable nullification of Stephenson & Dodds' patent. Edward Pease skilfully manipulated the image of Stephenson, recommending he 'should always be a gentleman in his dress' and speech, and that he appear smart and clean every day. Thanks to his friendship with the Peases he was able to access the friendly, entrepreneurial network of Quaker financiers, and through Revd Turner had a connection to the socially elite and technically minded Unitarians, whose influence far outstripped their numbers. But, ultimately, thanks to Nicholas Wood and Edward Pease promoting the man and the machine, George became known outside his native North East not just as a locomotive builder but as a railway builder – a crucial step on his way to building the Stockton & Darlington and then the Liverpool & Manchester Railway. He certainly understood the value of self-promotion, and before Samuel Smiles, the Victorian guru of self-help, had made him into the epitome of the self-made man George was a great proponent of that philosophy. He said:

Learn for thyself – think for thyself – make thine own answer – make thineself master of principles – persevere – be industrious – and then there is no fear of thee.

George Stephenson, 'father of the railways', was engineer of the Manchester & Leeds Railway, Leeds & Selby Railway, North Midland Railway (Derby to Leeds), Leeds & Bradford, and York &

Edward Pease, the Quaker businessman and financier behind the Stockton & Darlington Railway, did much to manipulate and promote the image of George Stephenson – both man and machine.

North Midland Railway (Leeds to York). In 1847 he was present at the meeting that led to the formation of the Institute of Mechanical Engineers and served as its first president. George was also a keen educationalist and supporter of the Mechanics' Institute movement, which was to provide a secular education for working-class men. Together with Revd William Turner he was important in establishing the Newcastle Mechanics' Institute and chaired its first meeting in 1824.

George ended his days a very wealthy man. He moved first to Alton Grange, and later to Tapton House near Chesterfield to better superintend the construction of the North Midland Railway and the Clay Cross Company. At Tapton he also indulged in his hobby of stockbreeding and growing monster vegetables and, inter alia, developing the straight cucumber. George married thrice; his second wife Elizabeth died in 1845 and he married again in 1848 to his housekeeper. He died in August 1848 and is buried at Chesterfield Parish Church, where he is commemorated in a stained-glass window.

To have risen the way George did must have taken serious strength of character, of will and drive. George was probably quite ruthless and while his attendance at Hanover Square Chapel, for example, may have been due to his own religious belief, it may also have been because Hanover Square and its minister represented the meeting place of the 'great and good' of Newcastle, particularly its engineers and entrepreneurs, giving him an inroad to their society and networks. He gave Robert the best possible start in life, but also appears to have lived through Robert – getting Robert an education in one of the best schools was one way of opening doors that had hitherto been closed to George. Having risen to such prominence, George was perhaps always conscious of his upbringing and lack of formal education. Certainly, his deep-seated mistrust of the London and 'establishment experts' was not just down to his being a dour, dissenting north countryman. George showed considerable drive and determination in pursuing the locomotive, and a locomotive-worked railway, when others had fallen by the way. He also had a vision not just of local lines but of a national network: when asked why the Canterbury & Whitstable had been built to the same gauge as the Stockton & Darlington and Liverpool & Manchester he replied it was because one day they would be connected. While perhaps not *the* father of the railways, he was certainly *a* father.

Commemorative plaque on the Stephenson cottage, unveiled in 1929 by the Lord Mayor of Newcastle to mark the centenary of the Rainhill Trials. (Lauren Jaye Gradwell)

CHAPTER 3

Robert Stephenson

Robert was George's only son with his first wife Fanny Henderson, born in October 1803. She was twelve years his senior and had been in domestic service when they first met. A daughter was born in 1805, but died at only three weeks old. Fanny, who had been suffering from tuberculosis, leading to George doing most of the child-rearing at home, succumbed to her illness in May 1806. Thereafter, George's sister, Eleanor, moved into the home to look after the young Robert.

Determined to give his son the best start in life, Robert was educated locally and then sent to the Dissenting Academy on Percy Street, Newcastle, run by John Bruce. The Percy Street Academy had been established as the law prevented Unitarians in particular from attending parish schools, and even forbade non-Anglicans from attending English universities. Studying at Percy Street involved a 10-mile walk each way, every day, for his lessons, so his father bought him a donkey on which to make the journey instead. The Percy Street Academy was perhaps the best school in Newcastle, and it was while studying there that he came to the attention of Revd William Turner (see Chapter 2). Robert would be forever grateful to Turner for the polish he gave his education:

Robert Stephenson, photographed later in life. The only son of George Stephenson, he grew up alongside the locomotive and together with Brunel and Joseph Locke formed the great triumvirate of Victorian engineers.

DIAL OVER THE DOOR OF STEPHENSON'S COTTAGE, AT KILLINGWORTH.

Above left and above right: Robert was born at Dial Cottage at West Moor, Killingworth, in 1803, so named after the sun dial he and his father made and placed over the door in 1816.

> Mr Turner was always ready to assist me with books, with instruments, and with counsel, gratuitously and cheerfully. He gave me the most valuable assistance and instruction; and to my dying day I can never forget the obligations which I owe to my venerable friend.

Turner was certainly a major influence on Robert's life. Another of Turner's pupils was Michael Longridge, who was destined to become a partner in Robert Stephenson & Co. While there is no record of the Stephensons being members of Turner's influential congregation at Hanover Square Chapel, they certainly met and became friendly with prominent Newcastle Unitarians who were members. It is tempting to see George in the pews at Hanover Square out of religious conviction, but it was also as a means of social climbing. Robert's education at Percy Street would have included theology, natural philosophy, history, geography, science, modern languages, and classics, all with that unique Unitarian stance of freedom, reason and tolerance. As such, Robert would have been encouraged to think for himself, to 'think outside the box'.

Thanks to Revd Turner, Robert became a member of the Newcastle Literary & Philosophical Society, which opened many doors for him. In 1819 Robert was apprenticed to Nicholas Wood to learn the trade of a colliery viewer and helped with the survey of the Stockton & Darlington Railway (1821). He then attended the University of Edinburgh for six months; this was a natural route for English dissenters who, denied a university education south of the border, looked to Edinburgh or Glasgow Universities to finish their education – there was no religious test for admission. Here he studied natural philosophy (although Robert had little patience for the subject), chemistry, and natural history, thoroughly enjoying the geology field trips.

George was what we might now term a pushy parent, on the one hand giving Robert the best possible start in life, but on the other using Robert to overcome his own shortcomings.

As a young man Robert came to the attention of Revd William Turner, the influential minister of Hanover Square Unitarian Chapel, Newcastle.

Robert became the technical and intellectual crutch upon which George relied due to the gaps in his own education. And George foundered in Robert's absence, especially on the Liverpool & Manchester contract and with the technical and managerial side of things at the Forth Street Works of Robert Stephenson & Co. (see pp. 32–36). It must have been quite claustrophobic for Robert, going to his lessons during the day and, as Smiles records, teaching his father in the evening. No wonder he broke with his father and travelled to South America to come out from under George's shadow and 'find himself' as an independent adult rather than the son of the famous father. Indeed, while George's character is easily identified and engaging, that of Robert is harder to identify and define.

Robert Stephenson & Co.

The Stockton & Darlington Railway was perhaps one of the most significant railways of the nineteenth century. While a public railway with capital held in shares and open to all users upon payment of a toll was not a new idea (the Lake Lock Rail Road in Yorkshire of 1796 is considered to be the earliest known example), nor locomotive haulage (Middleton, 1812) nor the carriage of passengers and the use of a timetable (Swansea & Mumbles, 1807), the S&D brought these ideas together. As such, it would influence all those railways that would come later, including the first mainline railway, the Liverpool & Manchester. The Stockton & Darlington was developed by Quaker financiers, the Peases of Darlington. Edward Pease (1767–1858) believed George Stephenson was the right man for the job to build this railway. Moreover, as the most experienced locomotive builder in the country, the perfect choice to design and construct its motive power. In 1821 George had entered into a partnership with John and Isaac Burrel of Newcastle, and it was they who built the iron bridge over the River Gaunless, which can now be seen at Locomotion Museum. Two years later a new partnership was formed: Robert Stephenson & Co. as engineers, millwrights and engine builders. Premises were built at Forth Street. The capital of £4,000 was divided into ten shares:

Edward Pease	Four shares
George Stephenson	Two shares
Robert Stephenson	Two shares
Michael Longridge	Two Shares

Robert, not yet twenty-one, was appointed as managing director, on a salary of £200, 'to take general charge of the manufactory' with his father 'furnish[ing] the Plans &c., which may be required'. Robert was 'called upon to superintend ... operations ... engage men, take orders, advise on contracts, draw plans, make estimates, keep the accounts, and in all matters great or small govern the ... establishment'.

A second partnership was formed in 1824: George Stephenson & Son, which had the same shareholders as Stephenson & Co., for civil engineering and railway contracting. Among the staff were many future railway engineers, including Joseph Locke who together with Robert Stephenson and Isambard Kingdom Brunel (1806–59) would be the great triumvirate of mid-nineteenth-century engineer. Often considered rivals, the three were actually firm friends, especially Stephenson and Brunel.

The young firm of Robert Stephenson & Co. soon got into difficulty. Early in 1824 Robert broke with his father to work for the Colombian Mining Association for three years, and George was away from Newcastle surveying the Liverpool & Manchester Railway. Thus, there was not only a lack of management but also technical expertise. The bulk of day-to-day management of the firm fell onto Michael Longridge, who was also managing the Bedlington Ironworks. James Kennedy (1797–1886), a Scottish engine-wright, was appointed as foreman for a period of eighteen months, and he was joined by Timothy Hackworth (see Chapter 6), then foreman-smith at Walbottle Colliery, who 'was able to bring well-founded experience to the Stephenson Concern' but left by the end of 1824. Kennedy, who is said to have introduced outside coupling rods during his time at Forth Street, would

Robert Stephenson & Co. were established on Forth Street, Newcastle, in 1824. These are the oldest surviving buildings, dated c. 1840. (Rob Langham)

go on to be a partner of Edward Bury (1794–1858) of the Clarence Foundry, Liverpool, the major rival to Robert Stephenson & Co. during the early 1830s (see Chapter 7).

It is little wonder that orders for the Stockton & Darlington Railway were delayed and that the first two locomotives, of rather rough construction, suffered from mechanical faults. Despite trying to get an early release from his agreement with the Colombian Mining Association, Robert returned to Britain in late November 1827. In Colombia he had met a penniless Richard Trevithick who, too, had been lured to South America by the promises of riches. Robert paid for Trevithick's voyage home, during which the pair were shipwrecked.

ENGINE USED IN CONSTRUCTING THE LIVERPOOL & MANCHESTER RAILWAY, 1828–9.

Despite the nonsense caption, this is Springwell No. 2, one of two locomotives built by Robert Stephenson & Co. in 1826 for the Springwell Colliery. No. 2 was photographed in 1863, and it remained at work until 1879. The Springwell locomotives were ordered before those for the Stockton & Darlington, but the order was delayed.

Above left and above right: *Locomotion* was one of four locomotives built in 1825 for the Stockton & Darlington Railway. Rebuilt several times during her working life and later loaned to the Pease family to work as a colliery pumping engine, she was restored in 1855 and put on display the following year. Perhaps uniquely, she remained in railway ownership until becoming part of the National Collection in 1975.

Having spent Christmas 1827 in London, he returned to Newcastle in January 1828 and set out design parameters for what would be a frenetic thirty-three months of work at Forth Street: to reduce the 'ugliness' of locomotives; to simplify their working gear; to make them easy to operate; and to place the cylinders 'either side of the boiler, or beneath it entirely' like the road coaches of Goldsworthy Gurney and with direct drive from the pistons to the wheels. Placing cylinders under the boiler required the use of a cranked axle, a major technological challenge. George had proposed their use in his 1815 patent, but forging crank axles by hand before the invention of the steam hammer was always done with difficulty.

Direct transmission from piston to wheel greatly improved efficiency and simplified construction and maintenance; suspension was improved through the use of steel leaf springs, an innovation of Nicholas Wood; the use of larger diameter wheels (another suggestion of Wood); hammer blow on the track was reduced by using inclined rather than vertical cylinders; and more economical use of steam via expansive working, by letting the steam expand in the cylinder. Experiments also took place on boiler efficiency, including increasing the number of boiler flues, and with burning alternative fuels such as coke (a relatively smokeless fuel) because the Liverpool & Manchester Railway Act forbade locomotives from making smoke. *Rocket*, the firm's entry for the Rainhill Trials, incorporated many lessons learned during this period: inclined cylinders with direct drive to the wheels, which were fully sprung; a multi-tubular boiler with a proper firebox; and a simple to use valve gear known as the 'flying reverse'. But *Rocket* was not the culminant point of this process. It was *Planet*, completed only twelve months after *Rocket*, which was the first 'modern' locomotive, having a multi-tubular boiler with a smokebox at one end and a firebox within the

The oldest known depiction of *Rocket*, drawn by Stephenson's rival, Charles Blacker Vignoles, for the *Mechanics' Magazine* of 24 October 1829.

boiler shell at the opposite, proper frames, was fully sprung and used horn guides, and had horizontal cylinders beneath the smokebox and a crank axle. It was a design that one of the directors of the Liverpool & Manchester Railway declared was as 'near to perfection' as it was then possible to be.

Planet, delivered in October 1830, was the culmination of Robert's frenetic thirty-three-month design and research project, which began in January 1828. This replica was built in 1992. (Will High)

The Planet class evolved to meet the operational demands of the Liverpool & Manchester Railway, with the 0-4-0 Samson being delivered in spring 1831 for working goods trains and banking duties. Then, in order to reduce axle load on the lightly laid track, a third axle was added, creating the Patentee type of 1833 for which Stephenson obtained a patent in that year. Patentees were built as a 2-2-2s for fast passenger business, 0-4-2s for goods trains, and as heavy-weight 0-6-0s for working heavy coal trains – the first of countless thousands of inside cylinder 0-6-0 goods engines on Britain's railways. His second patent design of 1841 was for the 'long boiler' locomotive, which was designed to give as long a boiler as possible (so that as much heat as possible was extracted from the combustion products), but with all the axles between the smokebox and firebox. Sadly, they had a reputation for bad riding at speed due to their short wheelbase and, from 1843, the use of outside cylinders. This rough riding, however, did lead to French engineer Louis Le Chatelier to carry out the first study of rail vehicle dynamics and introduce balance weights to wheels to counteract the effect of the pistons' connecting rods, etc. The type found great favour in France where the design took on a life of its own, evolving into the Polonceau and Bourbonais types.

Lion, the only surviving Liverpool & Manchester Railway locomotive, was built in Leeds by Todd, Kitson & Laird in 1838 to a design by Robert Stephenson. Seen here at Liverpool Road station, Manchester, 14 September 1980.

Railway Builder

Robert was not just a builder of locomotives but railways too. He was elected a member of the Institute of Civil Engineers in 1830 and was Chief Engineer of the London & Birmingham Railway, which finally opened between London Euston and Birmingham in 1838. As engineer of the Newcastle & Berwick Railway, Robert was responsible for the famous high-level bridge in Newcastle, which was formally opened by Queen Victoria in September 1849, realising his father's dream of a line of railway from Thames to Tyne.

Robert also engineered the Chester & Holyhead Railway, which involved the construction of two tubular bridges over the Conwy and the Menai Straits. In this he was assisted by his father's old friend, William Fairbairn, and Eaton Hodgkinson – the foremost authorities on iron bridges – and conducted exhaustive experiments using 1/6th scale model bridges, 78 feet long, at Fairbairn's ironworks to examine how a tubular bridge would perform. When the first of the great tubes of the Britannia Bridge were being raised into position in June 1849, Brunel (one of Robert's closest friends) was on hand. Robert placed the final rivet on 5 March 1850. The Britannia Bridge was considered a marvel of the age; sadly, it was destroyed by fire in 1970. Stephenson also built a lesser-known tubular bridge over the River Aire at Brotherton, which also opened in 1850. But it was on the Chester & Holyhead that Robert suffered one of his greatest setbacks when his bridge over the River Dee, near Chester, collapsed in May 1847. The bridge had collapsed due to the use of both cast and wrought iron in the same structure (known as a trussed compound girder) and following a royal commission into the disaster, the use of wrought-iron bridges became universal,

TUBULAR & SUSPENSION BRIDGES, CONWAY CASTLE. ERECTED 1850.

Robert was also a great bridge builder. He built the tubular bridge over the River Conwy at Conwy, running parallel to Thomas Telford's earlier suspension bridge.

One of Robert's greatest engineering triumphs was the tubular bridge built across the stormy Menai Straights. The magnificent structure was considered a wonder of the world when it was completed.

improving bridge safety. In 1859 he built the then longest bridge in the world, the Victoria Bridge over the St Lawrence River at Montreal.

Later Life

Together with Joseph Locke and Brunel, Robert Stephenson was part of the great triumvirate of Victorian engineers. He was a Fellow of the Royal Society; Hon. Fellow of the Royal Society of Edinburgh; member of the Institute of Civil Engineers; was honoured by French, Belgian and Swedish royalty; and built or was consulted on railways in Belgium, Denmark, Egypt, Italy and Switzerland. Like his father, he was also offered a knighthood but declined. He was the second president of the Institute of Mechanical Engineers upon the death of his father. He never seems to have left his father's shadow, even toward the end of his life, downplaying his own role in the evolution of the locomotive. Like Locke and Brunel, Robert was a workaholic who worked himself to an early grave. He had always suffered from depression, which was exacerbated following the death of his wife in 1842 and, like Brunel, became fond of cigars and opiates. Quite what his tutor Revd William Turner would have thought of Robert becoming Tory MP for Whitby on an economic protectionist card is not known, but his response would probably not have been favourable.

By 1850 Robert had been responsible for building a third of the country's railway system and was ill with chronic nephritis, a disease of the kidneys. For pleasure he took to yachting, and commissioned a 100-ton yacht christened *Titania*. He was with his friend Brunel during the troubled launch of the *Great Eastern*, but, falling on the slipway, he fell into the river and as a result of this drenching caught bronchitis. He sailed to Egypt on the *Titania* in 1858 and met Brunel for the last time in Cairo on Christmas Day. He was in Norway to receive the Knight Grand Cross of the Order of Saint Olaf in early autumn 1859, but fell ill at a banquet and had to return home. He died, aged just fifty-five, in October 1859. At his funeral in Westminster Abbey, 3,000 were admitted to the service and on the same day

THE "TITANIA" SCHOONER YACHT, BUILT FOR MR. ROBERT STEPHENSON, C.E.

Left: *Titania* was built as Robert's private yacht. For a workaholic like Stephenson, yachting provided a much-needed escape from the world of work.

Below: *Rocket* is perhaps Robert's most well-known legacy. The surviving components of the 192-year-old locomotive are seen here on display at the Science & Industry Museum, Manchester. (Ian Hardman)

1,500 employees of Robert Stephenson & Co. marched through the streets of Newcastle to their own memorial service at Newcastle Parish Church (Cathedral from 1882).

Robert had grown up with the locomotive; his father's first attempt, *Blucher*, had been built when he was ten years old and he and the locomotive grew to maturity alongside each other. At the age of twenty-one he was appointed managing director of the Robert Stephenson & Co. and was just shy of his twenty-sixth birthday at the Rainhill Trials. He was one of the foremost locomotive, railway and bridge engineers of his age.

CHAPTER 4

Henry Booth

Henry Booth, designer of *Rocket*'s boiler, was born in Liverpool in 1788. He was the eldest of five sons born to Thomas Booth, a wealthy Liverpool Unitarian corn merchant with premises on King Street. Father and uncle (George) were both Freemen of Liverpool, and the Booths became the unofficial spokesmen for the Liverpool corn trade.

The biggest influence on Henry's life was his Unitarian faith and education, first by Revd Robert Lewin (1739–1825) of Benn's Garden Chapel and then by Revd William Shepherd of Gateacre (1768–1847). Benn's Garden Chapel was:

> The meeting house for a tightly-knit network of Unitarian ship owners and merchants who frequently formed alliances by marriage, met socially, invested in one another's ventures, shared or exchanged practical skills, embarked on philanthropic (especially educational) schemes, and engaged fully in the politics of reform.

Henry Booth of Liverpool was the first 'modern' railway manager, serving as Secretary, Treasurer and General Manager of the Liverpool & Manchester Railway, later serving the London & North Western Railway until 1859.

Members of the well-connected congregation included William Roscoe (1753–1831, MP for Liverpool in 1806), William Rathbone IV (1757–1809) and William Rathbone V (1787–1868). Roscoe was a wealthy lawyer. In the early 1790s he got into the coal trade and in 1796 gave up his legal practice to drain and cultivate Chat Moss and Trafford Moss near Manchester. The Rathbones were originally a Quaker family and had risen to prominence in Liverpool society as merchants and shipowners. Roscoe and Rathbone were 'the mainstay of the reforming party' in Liverpool and in that part of Lancashire.

Revd Shepherd was a political radical, values that he would instil in the young Henry. Shepherd was an associate of the Unitarian Mary Wollstonecraft, who wrote *A Vindication of the Rights of Woman* in 1792, and he publicly supported female suffrage in 1794. 'A terrier in size and disposition ... His wit disguised his strong sense of political commitment.'

Henry was educated during the tumultuous years of the French Revolution by religious and political radicals who shared a fierce belief in freedom, reason and tolerance. Shepherd as well as Booth's father and uncle were part of a group of radicals that surrounded William Roscoe, known as the 'Roscoe Circle' or by their opponents the 'Liverpool Jacobins' from their support of the ideals of liberty, equality, and fraternity of the French Revolution. Benn's Garden Chapel was the nursery of 'civic radicalism' in Liverpool and its largely middle-class congregants were well educated, earnest, freethinkers (Unitarians reject all formal creeds) who embraced radical politics, championing the abolition of slavery (which didn't make them popular in a Liverpool dominated by what was euphemistically called the 'African Trade'), abolition of the Test and Corporation Act, disestablishment of the Church of England, religious freedom, and universal suffrage.

SACRED TO THE MEMORY OF
WILLIAM SHEPHERD LL D
FOR FIFTY-SIX YEARS MINISTER OF THIS CHAPEL
A MAN OF UNDEVIATING INTEGRITY IN ALL THE RELATIONS OF LIFE

Revd William Shepherd of Gateacre Unitarian Chapel was a major influence on Henry Booth, instilling in him radical political values including religious freedom, abolitionism and universal suffrage. (Clare Grace Williamson/Gateacre Unitarian)

The Roscoe Circle was dominated by Unitarians and included William Rathbone IV; Revd William Shepherd; Revd John Yates (1755–1826) and Revd Joseph Smith (who was co-pastor at Benn's Garden Chapel with Revd Lewin); Dr James Currie FRS (1756–1805), the physician at Liverpool Infirmary whose son, William Wallace Currie (1784–1840), would be the first mayor of Liverpool in 1836 and a board member of the Liverpool & Manchester Railway; and the Booth brothers. Roscoe also attracted several prominent Quaker families, such as the Croppers – most notably James Cropper (1773–1840) a leading abolitionist and a future director of the Liverpool & Manchester Railway. They were all united, however, in their opposition to war with France, and as F. E. Sanderson has stated: 'Without exception, the members of the Roscoe Circle welcomed the news of the French Revolution as the herald of better times ahead for ... Reformers and Dissenters.'

The greatest triumph of the Roscoe Circle was the election of Roscoe as MP for Liverpool in 1806, in which Thomas and George Booth played an important role. Roscoe was therefore able to vote for the act which abolished slavery. But in Liverpool there were riotous scenes as an angry mob vented its frustration against Roscoe and his abolitionist supporters. He was unseated in 1811 because of his abolitionist views.

Support for the ideals of liberty and equality, at a time when Britain was at war with Republican France founded upon those values, made the Unitarians dangerous and victims of sectarian violence known as the 'English Reign of Terror' – the result of William Pitt repealing the right of Habbeus Corpus in 1792. This led to angry mobs attacking

The pulpit and organ at Renshaw Street Unitarian Chapel, Liverpool, where the Booth family were worshippers alongside other families of the 'great and good' in the town.

LIVERPOOL.
RENSHAW STREET CHAPEL PULPIT.
1811.

43

known Unitarians and their places of worship in Manchester, Liverpool, Nottingham and elsewhere, including the near murder of Revd Dr Joseph Priestley (1733–1804) in Birmingham. Priestley fled to America, and in Liverpool William Roscoe, Revd William Shepherd and Dr Currie made similar preparations to flee, but: 'During the dark years of the war, the small body of English Unitarians ... played an astonishingly large part in keeping, at the constant risk of violence, or of imprisonment, thought and the hope of progress alive.' (Holt, p. 121.)

Pamphlets and Politics

Henry completed his studies with Revd Shepherd in 1804 and sometime before 1811 he became the Liverpool agent for the Albion Fire & Life Insurance Company, continuing in this role until 1816 when his younger brother George took over.

In 1811 the old Benn's Garden Chapel was replaced by a new building on Renshaw Street, to which the Booths gave handsomely. The chapel was opened on 20 October 1811, and Henry was consulted about (and probably designed) the hot water heating system, installed in 1822. In August 1812 he married Ellen Crompton (1789–1871), daughter of Abraham Crompton (1753–1829) of Chorley Hall and a Unitarian. Ellen's sister, Jessy, married into the wealthy Potter family of Glossop and was thus related to Beatrix Potter, the writer.

It should be no surprise that Henry was interested in radical politics. As Smiles wrote: 'Henry Booth was of a nature and temperament which would not permit him to live in close contact with a felt grievance and injustice without strenuous and persistent protest and remonstrance against it.' (Smiles, p. 79.)

He joined the Liverpool Literary & Philosophical Society in 1813. His first published political essay was 'Moral Capability', which was published the following year in the Unitarian *Monthly Repository of Theology and General Literature*, putting him on the national stage for the first time. Also in 1813 he joined the Liverpool Concentric Society, a reform-minded organisation, the first president of which was Colonel George Williams. It was a predominantly middle-class body but numbered 1,000 members calling for an end to the costly war with France, universal suffrage, and vote by private ballot. The years leading up to and after Waterloo were turbulent: there had been consecutive bad harvests leading to food shortages, a banking crisis, and mechanisation was causing a high rate of unemployment. There was civil unrest and the clamour for reform came even louder, which the government of the day tried to quash via the 'Gag Acts' (1811–16), and in 1817 the 'Four Acts', which included the repeal of Habbeus Corpus. It should not be surprising that Henry Booth, his father, Revd William Shepherd and Joseph Sandars (1785–1860, a fellow Unitarian, corn merchant, and member of the Concentric Society) would be present, and speak, at a meeting held in Liverpool to urge the government 'against the further suspension of the Habbeus Corpus Act'. Two years later, members of the Concentric Society were invited to speak at the mass reform meeting in Manchester on 16 August 1819, which would sadly end with the Peterloo Massacre. Henry and Thomas Booth, Revd William Shepherd, and Joseph Sandars would lend their names to a petition to the Prince Regent for an inquiry into the events in Manchester (which was denied) and also donate money for those who had been killed or wounded.

Members of the Liverpool Concentric Society, which included Henry Booth and Joseph Sandars, were invited to speak at the peaceful political rally in Manchester on 16 August 1819. Panicky magistrates ordered troops to disperse the crowd, resulting in the Peterloo Massacre. Henry would add his name to a petition for an investigation into the massacre – which was denied.

Booth and his fellow reformers were also interested in global events, holding a public meeting in support of Spanish liberals in the face of military intervention by the French to restore traditional rule (1822–23) and showing support of Simon Bolivar and his liberation of Bolivia from Spanish rule (1825). He also spoke in favour of the Greeks in their fight for independence from Turkey (1821–29). In summer 1830, when Booth would have been busy with the final arrangements for the opening of the Liverpool & Manchester Railway, Charles X of France attempted a *coup d'etat* to establish himself as an absolute monarch, leading to a three-day revolution in Paris that saw the French crown pass from the older Bourbon line to the Orléans dynasty. In Liverpool, a public meeting was held in favour of the liberal, new French king, at which Booth and Revd Shepherd spoke. Both subscribed to a fund for the benefit of the families of the dead and wounded Parisians who had taken to the barricades against the tyrannical Charles X.

At home, Booth wrote pamphlets on the condition of the urban poor, linking poverty and overpopulation to lack of birth control. He also campaigned against the Corn Laws, which set domestic corn prices artificially high and restricted the import of cheap corn from Europe or the Baltic. As a Unitarian he also supported the repeal of the Test & Corporation Act, which pretty much rendered null the rights of non-Anglicans in England. It was partially repealed in 1828 and Booth and his tutor, Shepherd, were present at a celebratory banquet in Liverpool. Booth also publicly spoke in favour of the Civil Marriage Bill, the passage of which meant that non-Anglicans could get married according to their own rites and customs. He also supported the Dissenters' Chapel Act (1844), which gave Unitarian congregations the right to places of worship. Booth was always on what now would be termed the 'left' of politics: he was a vocal champion of the 1832 Reform Act, making several public speeches and attacking in particular the then prime minister, the Duke of Wellington and the 'corrupt cronyism of Toryism'.

Railway Promoter

Booth had been raised to believe in progress, and he was also part of a well-connected and influential network of wealthy businessmen who represented the elite of Nonconformity in Liverpool. As A. H. John has noted, the 'collective influence of these families on the trade of Liverpool' was enormous. Unorthodox in theology but uncompromising in their ethics, they were cultured yet radical in their political and religious outlook, and men like Henry Booth were firmly planted in a well-connected and often intermarried network of men (and women) who believed in progress. Booth had the vision, influence and money to support the idea of a railway between Liverpool and Manchester.

Liverpool corn merchants were bitterly complaining of the inadequacy of the existing transport from Liverpool to Manchester by road, but especially by canal, when a cargo could take several weeks to travel the 30 miles. William James (1771–1837) surveyed and proposed a tramway between the two towns in 1821 and the idea was enthusiastically taken up by Joseph Sandars (see p. 44). The second survey was completed in 1822 and a

The civil engineer William James first had the idea for a railway between Liverpool and Manchester, and carried out the first survey of the line in 1822.

provisional committee was established, of which Henry Booth was a member from 1824. In May 1824, Booth, together with Joseph Sandars, John Kennedy (see Chapter 1) and Lister Ellis, visited all the different railways in the north of England to study their workings and report their findings to the committee. They wholeheartedly reported in favour of locomotive traction.

Booth 'quickly became second only to Sandars in working for the railway project ... [and] ... demonstrated an unusual gift for promotional and organisational work; from the beginning he gave a new vitality to the project.' He was a skilful and energetic organiser with 'remarkable administrative ability' and it was these various talents that were immediately recognised by the committee, who appointed him their secretary.

The L&M Railway came into being on 24 May 1824 and a prospectus was issued in July. Among the subscribers to the company were Thomas, George and Henry Booth. In order to form a company and to get the railway built, the committee would need an Act of Parliament. The company submitted its first bill to Parliament in 1824, but was thrown out in 1825 due to the errors in George Stephenson's survey and his poor performance before a Parliamentary committee. The Railway Committee met to lick its wounds and a second prospectus was issued on Boxing Day in 1825, and the second bill became law on 5 May 1826. The proprietors of the new company met on 29 May 1826, chaired by Charles Lawrence, at which Booth was appointed as treasurer and secretary on a salary of £500 per annum. The first board meeting was held the following day, and from then on met every Monday at 12 noon; each director was paid 1 guinea per meeting attended.

Painted by Spiridone Gambardella in 1853 and presented to Joseph Sandars, this depicts Sandars (centre) in conversation with George Stephenson (right) and Charles Sylsvester (left – an early proponent of railways), discussing the Liverpool & Manchester Railway. The painting is now lost. (Liverpool & Manchester Railway Trust)

Railway Manager

Henry Booth was the first 'modern' railway manager. He had taken a leading role in formation of the L&M Company and was its secretary and treasurer. Not only did the L&M pioneer a steam-worked passenger and goods service worked to a timetable, but railway management too. Earlier public railways like the S&D had a poorly defined management structure; few salaried, professional managers; and placed great reliance upon subcontracting 'if not as a method of management, at least as a method of evading management'. The organisation of the L&M into specific operating departments – Engineers (encompassing civil and mechanical engineering and motive power); Coaching (the passenger business); Carrying (the goods business); Minerals (coal and coke) – each with its own salaried, professional manager was one that would be adopted by all those railways that followed in its wake. However, following the dismissal of Anthony Harding (the locomotive foreman) and his brothers for gross misconduct, and an inquiry into alleged accounting deficiencies in June 1833 the board decided that Henry Booth's de facto position be legitimised and given an even broader remit, assuming the 'General Superintendence of the various Departments of the Concern, should have power to decide & act on all points arising out of the ordinary operations of the Road and touching the general business of the Company'.

Through his 'close, active, and unremitting personal superintendence' Booth was responsible for the general organisation, discipline and welfare of the company on a salary of £1,500 per year. And, in order that Booth 'be more at liberty to exercise a personal Superintendence & Control over the line of Railway & at the various Stations' a team of four (later six) clerks was taken on at Liverpool. These clerks would be responsible for the 'principal part of the office work' that hitherto had been done by Booth, including 'the ordinary correspondence, the arrangement of papers, minutes of proceedings &c'. Such a small number of clerks to run such a large concern as the L&M should not be surprising: 'Offices at this time ... comprised small groups of clerks and apprentices, all of whom worked in close contact with their principals and many of whom were drawn from the same social strata.' (John, p. 159.)

The 'hours of work were long but the pace leisurely', and there was a close relationship between staff in such offices, with high degrees of trust and personal responsibility. In order to relieve the board of directors from the rigours of day-to-day management of the railway, a Management Committee was established in 1831, which met weekly on Thursdays (pay day), and Booth was responsible to both. With the appointment of Booth as general superintendent, the committee met fortnightly from July 1833. Members were at first unpaid until 1836 when they drew 6 guineas a fortnight to be shared among those who attended.

The board at first had a paternalistic attitude toward their employees. If any member of staff was injured or unable to work due to illness they received half pay, and the company would pay for medical attendance. The board also approved the establishment of an 'Annuity Fund' in 1831 for all railway staff 'to be Supported by a small weekly contribution out of the Wages of the Men', which would pay out to those men too sick to work or injured in the line of duty. It also paid out to widows and orphans. Thereafter, however, attitudes became more laissez-faire – not least the result of the first railway strike in February 1836, which Booth had moved quickly to suppress. Trade unions, or 'combinations', and strikes were then illegal and even the liberal Unitarian minister Revd William Gaskell of Manchester described both as 'unnatural'. Ultimately, four of the turnouts were sentenced to four months' hard labour on the treadmill at Kirkdale Gaol, but after receiving a letter from the gaol chaplain the board agreed to have the sentence commuted to imprisonment only. Sick pay was abolished in 1837, and instead staff were encouraged to enrol in a 'Friendly Society', as the board was reluctant to pay for a doctor or surgeon's fee for medical care. By 1841 the board 'consider[ed] themselves absolved from any obligation to assist men under accidents or sickness' who were not members of a benefit club. That said, in August 1843 the board established a 'Reading Room' for the benefit of company employees, which had a lending library, reading room stocked with periodicals and an organised series of 'Mutual Improvement Classes'. By the 1850s Booth's attitude towards trade unions had softened, and he urged the London & North Western Railway to consider some 'legislative sanction for mutual and reciprocal engagements ... between Railway Companies and Enginemen as may prevent ... [a] ... strike among Enginemen'.

The L&M merged with the Grand Junction Railway in 1845 and in the following year the London & North Western Railway came into existence, and Booth was voted a 'handsome purse' of 3,000 guineas (about £70,000 today). He would remain as secretary to the Northern Division of the LNWR, and also served on the Northern or Crewe Committee and Northern Locomotive Committee.

Patents and Inventions

Henry had received a wide and liberal education, which taught him to think for himself and to 'think outside the box.' He had a mechanical turn of mind and at home his study was set up for 'scientific and mechanical experiments' and was where he 'planned and matured his innovations'. Between 1835 and 1837 he obtained five patents to do with railway locomotives or carriages and the railway system. His first patent was a design for a steam boat (1819). Steam boats had been introduced on the Mersey in 1815, and in 1821 Booth became part-owner of the *Cambria*, the first steam packet to carry passengers from Liverpool to North Wales. His 1819 patent (No. 4367) was for propelling canal boats by means of a reciprocating frame fitted with hinged boards, which moved backwards and forwards rather like a mechanical oar, powered by a steam engine. It was probably only ever a theoretical idea as it cannot have been very efficient. As noted earlier, Booth also designed the hot water heating system for Renshaw Street Chapel in 1822.

Perhaps his most famous collaboration was with George and Robert Stephenson on coke burning boilers. Hitherto locomotives had burned coal, but because the Liverpool & Manchester Railway Act prohibited locomotives from making smoke an alternative, smokeless, fuel was needed. Coke is made by heating coal in a retort to drive off the volatile hydrocarbons, which produce black smoke if not properly burned. Thus, coke is nearly pure carbon, and in order to burn cleanly and efficiently requires a large amount of oxygen.

Earlier locomotive boilers had either a single, large-diameter flue running straight through the boiler or a return flue, which, although it greatly increased the heating surface, placed the chimney somewhat awkwardly at the same end of the boiler as the firebox. Booth's first tentative designs for what would be a multi-tubular boiler were made in association with George Stephenson in 1827–28. In April 1827 Booth reported to the L&M that he had had idea of how to burn effectively burn coke in a locomotive boiler. The board approved George Stephenson and Booth to carry out experiments with a coke-burning boiler, and granted £100 to finance the work.

Stephenson reported to the board in January 1828 that the experiments had been successful, and a drawing of the proposed new boiler and locomotive was also exhibited to the board, who agreed that a locomotive should be built and must cost no more than £550. This was the *Lancashire Witch*, a compact 0-4-0 and the first Stephenson locomotive to have direct-drive from the pistons to the wheels; it was carried on springs and the steam could be worked expansively. By burning coke they had mitigated the nuisance of smoke, and Stephenson and Booth set about removing another: noise. As Nicholas Wood had described, exhaust steam directed up the locomotive's chimney made a 'very disagreeble noise' and one that many influential landowners objected to. To this end the blast characteristic was softened as much as practicable and instead compressed air, provided by bellows under the tender and driven by eccentrics from the

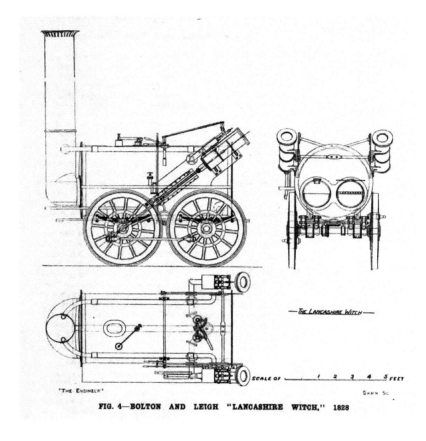

— The Lancashire Witch —

SCALE OF ___ 1 2 3 4 5 FEET

"THE ENGINEER"   SWAIN SC.

FIG. 4—BOLTON AND LEIGH "LANCASHIRE WITCH," 1828

Lancashire Witch, built by Robert Stephenson & Co., incorporated a boiler designed by Henry Booth. At first it had a double-return flue, so that the main boiler flue bifurcated and returned as two smaller flues, terminating in their own chimney. In rebuilt form it had a breaches tube or Y-shaped flue with two separate fireboxes.

Booth's second boiler design was that for *Twin Sisters*, an unusual twin-boilered, twin-chimenyed locomotive used in the construction of the Liverpool & Manchester. (Liverpool & Manchester Railway Trust)

wheels, was utilised to liven the fire. As George Stephenson wrote to Timothy Hackworth this worked well and importantly there was 'not the least noise' about it. Bellows were soon dispensed with, however, as it was found working them absorbed too much of the locomotive's power output. Ultimately, *Lancashire Witch* was never used on the L&M and was instead used by the neighbouring Bolton & Leigh Railway. Booth's second boiler design was that for the twin-boilered *Twin Sisters*, which was the first locomotive to be owned and used by the L&M, although never taken into stock. The boilers were built by Laird at Birkenhead.

His third boiler design was that for *Rocket*, which utilised twenty 3-inch-outside-diameter copper tubes to carry the hot gases through the boiler barrel, greatly increasing the heating surface (and thus steam-raising ability) compared to the early single- or return-flue boiler. The idea to build *Rocket* and enter her in the Rainhill Trials had been Booth's. George Stephenson mulled over the merits and joined with Booth in building *Rocket*. It was one thing to have suggested the design, but to actually construct such a boiler was another matter. It was carried out in Newcastle by Robert Stephenson, ably assisted by William and Ralph Hutchinson. It was for this reason that, following pressure from his father, Robert was later admitted into the partnership, each taking an equal share of the risk and the financial reward. Robert Stephenson and Booth were in

Above and below: Several replicas of *Rocket* have been built down the years – the first by the LNWR at Crewe in 1881. Robert Stephenson & Hawthorn of Darlington built a working replica for Henry Ford in 1929, the first of four such replicas built between 1929 and 1935.

regular correspondence during the making of the boiler, with Robert reporting several problems in keeping the boiler water-tight and stopping the ends deflecting during hydraulic testing to three times its working pressure as stated in the stipulations for the Rainhill Trials. The use of a multi-tubular boiler was not new of course, the first practical essay having been that of the Marquis Jouffroy d'Abbans on his steam boat in the 1780s and the first engineer to have the idea of applying a multi-tubular boiler to a railway locomotive was another Frenchman, Marc Séguin (1786–1875), over a year before *Rocket* was thought about.

Booth's other contributions to the development of railways included sprung draw-and-buffing gear using a complicated system of rods and leaf springs; the screw coupling (both 1836); and a type of axle grease (1835). He also patented a form of firebox to prevent the emission of smoke from coal but there is no evidence one was ever constructed. A patent of December 1836 referred to the safe working of locomotives through railway tunnels. Booth was also interested in rails, and developed an asymmetrical rail that was used on the Liverpool & Manchester (1837).

A working replica of *Rocket* was built in a project led by Mike Satow in 1979–80 to take part in the 'Rocket 150' event at Bold Colliery in May 1980. It used some components from the LNWR replica.

Above: The 1979 replica was rebuilt in 2009 with a new, more accurate boiler and firebox and she continues to give joy to many thousands of visitors to the National Railway Museum, York. Seen here with a train of period rolling stock. (Beth Furness)

Left: Driver's-eye view of *Rocket*. (Beth Furness)

The idea of a multi-tubular boiler was not new in 1829; the first engineer who had the idea of applying one to a locomotive was Marc Séguin in France over a year before *Rocket* was steamed.

Above and right: A life-sized statue of Henry Booth was erected at St George's Hall, Liverpool. He is depicted with his patented three-link screw coupling, as well as a drawing of *Rocket* for which he designed the boiler. (Liverpool & Manchester Railway Trust)

Later Career

Booth was secretary and treasurer of the L&M for its entire existence, and it was a position he would retain following its merger with the Grand Junction Railway (1845) and formation of the London & North Western Railway in 1846. He finally retired from the LNWR in May 1859, the 'father of railway management'. It is also thanks to the efforts of Henry Booth that Greenwich Mean Time was established as universal across the UK's railway network in 1844–47, but it was only finally adopted as Standard Time in 1880 in mainland Britain. He was a leading light of the first Railway Conference of 1841, which proposed, and adopted, standard railway signalling based on the rules of the Liverpool & Manchester. He continued to take a keen interest in railway safety to the end of his life, and it is thanks to Booth that the magnificent Huddersfield station was not partially demolished in 1850 when the LNWR Board deemed it unjustifiably splendid and expensive. Booth also continued as a pamphleteer, including one supporting compensation for victims of railway accidents, about labour law and the relationship between capital and labour. Booth died on Easter Day 1869 and is buried at the Ancient Chapel of Toxteth. A memorial tablet was erected at Renshaw Street Chapel, later removed to the cathedral-like Ullet Road Unitarian Church. A life-sized statue of Booth proudly stands at St George's Hall, Liverpool.

Henry Booth's grave at the Ancient Chapel of Toxteth. He died on Easter Sunday, 1869, aged eighty, having lived a full and active life. (Liverpool & Manchester Railway Trust)

CHAPTER 5

The Two Johns

Novelty was built by the international duo of John Braithwaite of London and the Swedish army officer John Ericsson. In this venture they were also materially assisted by the Anglo-Irish engineer Charles Vignoles. Unlike George Stephenson or Timothy Hackworth, Braithwaite was part of the London 'establishment'. He was probably the type of 'London expert' who could easily ruffle George's feathers, who had denied his role in the creation of the miner's safety lamp and then that he could ever cross Chat Moss.

The English Engineer

John Braithwaite Jr (1797–1870) was the third son of the civil engineer John Braithwaite Sr (1760–1818) and Eliza (née Doyle), while the grandfather (William, 1732–1800) was also in the engineering or metalworking trade. The family business on the New Road, St Pancras,

John Braithwaite, the youthful London engineer who formed one half of 'Team Novelty' at Rainhill.

was mostly involved with well-sinking – something continued by the fourth son, Frederick (1798–1865). Together with his older brother William (1757–1802), John Braithwaite Sr developed the modern diving bell, it being used to recover wrecks such as HMS *Royal George*, which sank at her moorings in Portsmouth in 1782 with the loss of 800 lives.

John was educated at Lord's grammar school in Tooting and entered the family business, accompanying his father in the recovery of the entire cargo and £130,000 in coin from the East India merchantman *Abergavenny* in 1806–07. The proceeds from this recovery set the family up for life, with John Sr being able to purchase the Old Manor House at Westbourne Green. Tragedy struck the family in 1818 when John Sr was killed in a highway robbery, leaving the enormous sum of £30,000 to his children. The family engineering business was inherited by his sons John Jr and Francis (d. 1823). Under John's skilful management, the family firm expanded to designing and building steam engines, particularly for the London brewing trade and for waterworks. Engines were supplied as far afield as Scotland, Ireland and the West Indies. Like Trevithick before him and like Matthew Murray in Leeds (1765–1826), Braithwaite also began developing high-pressure steam engines. As a result of his understanding of high-pressure steam, he was appointed to report to Parliamentary Select Committee to investigate the boiler explosion on a steam boat in Norwich (1817) and he was also responsible for installing air pumps to ventilate the House of Commons in 1820.

The Swedish Officer

Johan Ericsson was the son of well-respected Swedish engineer Olof Ericsson (1778–1818) in 1803; his mother, Britta Sophia, also came from a long line of engineers and Olof had succeeded Britta's father (Johan Yngström) as head of the local iron mines. In 1810 Olof joined the team of engineers building the Göta Canal, under the direction of Count Baltzar von Platen, with labour coming from 7,000 men of the Mechanical Corps of the Swedish navy.

Often the smartest man in the room, Captain Johan Ericsson was co-designer of *Novelty*. Many of his designs, while theoretically brilliant, were not practical. Today he is best remembered as designer of USS *Monitor*.

Johan and his elder brother Nils (1802–70) were tutored privately with every intention of their joining their father as an engineer. In 1814 both were accepted as cadets of the Mechanical Corps of the Swedish navy. In the same year the school was visited by the prominent English civil engineer Thomas Telford, who had advised von Platen on his canal project. By the age of fourteen, Johan was the rising star of the school and had been responsible for making fair copies of all the technical drawings for the Göta Canal, and had 600 labourers under his command. Sadly, due to an economic downturn, the canal project was scaled back in 1817, and tragedy struck the Ericsson family when Olof died aged forty the following year.

Dissatisfied with service in the Swedish navy, he left the navy and the canal project to work as a surveyor and civil engineer, aged only seventeen. He was commissioned as ensign (the lowest officer rank) in July 1820 in the 23 Regiment of the Royal Rifle Corps – his previous mentor, von Platen, was deeply upset and told Ericsson to 'go to hell'. He was promoted to lieutenant in November 1821, but military service was not for the young lieutenant. He 'chafed at the bit' and he resented the drill and blind discipline. What Ericsson did enjoy, however, was exercises such as callisthenics and gymnastics, and wrestling. 'Tis said he hurt his back trying to lift a cannon barrel in a show of strength.

During 1822 he was sent to the north of Sweden as a topographical engineer, gaining a reputation as a capable map-maker and brilliant draughtsman, serving in the Swedish army making maps for the next four years. Nils too would join the army as an officer with the engineers, but later joined the Mechanical Corps of the Swedish navy, retiring with the rank of colonel. He had a distinguished career as a civil engineer, canal and railway builder. Such were the quality of Ericsson's maps that they caught the eye of King Karl XIV Johan (1763–1844, who before his ascension to the throne had been Maréchal Jean-Baptiste Bernadotte in the service of Napoleon I), who desired Ericsson to produce maps of his campaigns in France.

Ericsson turned his formidable intellect to matters mechanical in 1821 and designed and built a printing press and then in 1823 the first of many hot air engines, which he considered to be far superior to the steam engine as there was no boiler that might explode or steam that could scald. He applied for a patent for his invention in 1826. During this period Ericsson had a flirtatious love affair with Carolina Christina, a son being born out of wedlock in November 1824. By the mid-1820s Ericsson was seeking pastures new and, thanks to an introduction to the British ambassador at Stockholm and Count Adolphe Eugéne von Rosen (who too had served in the Mechanical Corps of the Swedish navy), he took twelve months leave from the army and set out for Britain. Unfortunately, while in Britain Ericsson inadvertently went absent without leave by remaining too long in London and away from the army. Because of his failure to return to Sweden in October 1827 Ericsson was considered to have deserted the army, but with the help of friends in Sweden and intervention by the king, he was reinstated in the army with the rank of captain and allowed to honourably resign.

Steam Fire Pumps

Johan Ericsson arrived in London in 1826 eager to demonstrate his hot air engine, for which he tried to obtain a patent. The engineering establishment, however, were sceptical of his machine and the idea proved to be a failure. But he did manage to take out a British patent in August 1826 under the name of his friend von Rosen; a second patent was obtained for a water pump, operated by compressed air in December 1826 under the name of his new London friend, Charles Seidler, a German immigrant.

At the same time Ericsson was making his tentative steps in London, John Braithwaite met both George and Robert Stephenson; however, it was at a demonstration of Ericsson's water pump he met and befriended the Swede. Braithwaite was an established, and respected, engineer and the pair soon became firm friends and business partners.

Braithwaite and Ericsson developed one of the earliest steam fire pumps. Their first essay – sharing its name with the later locomotive *Novelty* – was built in 1829 and used a highly efficient boiler designed, and later patented, by Braithwaite, Ericsson and Vignoles. This first fire engine was used to great success fighting a fire at the Argyll Rooms in London, managing to pump 30 to 40 tons of water an hour up to a height of 90 feet. It had two cylinders, 7 x 16 inches, producing 10 hp, and weighed just over 2 tons. It was carried, like its locomotive namesake, on wheels designed and patented by Theodore Jones of London in 1826. It was also used to fight fares at the English Opera House and at a brewery, but despite this success Braithwaite 'received little patronage and support from the general public' and found that the fire insurance companies who would have benefited from the invention were also nonplussed.

A second engine was built in 1831 with 7 x 18 inch cylinders and the same type of boiler as *Novelty*. It was demonstrated in France and in Russia 'with great success'. A third fire engine for the Mersey Docks & Harbour Board was completed in the same year. It had twin horizontal cylinders working a crankshaft, which drove three pumps and developed 15 hp. It cost £1,200. A fourth engine, *Comet,* was built in 1832 for the king of Prussia, with horizontal cylinders at 12 x 14 inches, and the boiler was pressed to 70 psi. Steam could be raised in under twenty minutes and was able to pump 90 tons of water an hour to a maximum height of 120 feet. The boiler was remarkably economical and burned only 3 bushels of coke per hour. A fifth, experimental, engine was built in 1833.

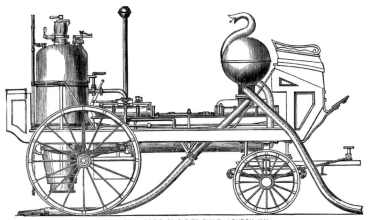

FIRST STEAM FIRE-ENGINE EVER BUILT—LONDON, 1829.
The " Novelty," George Braithwaite, Builder.

Braithwaite & Ericsson's first foray into steam power as a series of steam-powered fire pumps. The first of which, Novelty, was of a very similar design to – and may indeed have formed the basis of – the locomotive of the same name.

Unfortunately for Braithwaite and Ericsson the London Fire Engine Establishment was opposed to steam-driven fire pumps on land, believing there was insufficient water for them, and they met with 'frivolous objections and determined hostility':

> First, it was urged to be good for anything, it must constantly have a fire alight or the steam kept up ... then it was 'too powerful for common use, too heavy for rapid travelling, and requiring larger supplies of water than could be obtained in London streets'; and that even *if they could* get the water ... the quantity of water thrown *might* be 'injudiciously applied' and cause mischief!

The London Fire Engine Establishment, however, had no problems with a waterborne steam pump, and one was duly built by Braithwaite & Ericsson in 1835. It had an 80-foot-long iron hull, 13-foot beam and was 7 feet 4 inches from deck to keel. It was driven by a twin-cylinder engine (cylinders 16 x 22 inches), which were capable of both pumping water and driving the paddle wheels. It was considered to be the most powerful firefighting engine of its day.

This was not the end of the fire engine story. In 1839 Ericsson emigrated to the United States and in 1840, following a recent spate of fires in New York City, the Mechanic's Institute there offered a gold medal for the best design of steam fire engine; Ericsson duly entered and won. Steam was raised using a locomotive-type multi-tubular boiler with twenty-seven tubes, 1½ inches in diameter, and the fire was urged by bellows worked by the engine. Ericsson believed that by this means steam could be raised in ten minutes. It was thought capable of doing the work of 108 men and was able to pump around 3 tons of water an hour.

Steam Boats

Thanks to John Braithwaite's business connections, Ericsson became acquainted with Felix Booth, a gin maker who in 1828 introduced Captain John Scott Ross RN (1777–1856). Ross needed assistance with a steamship and, even better, the navy were willing to pay. The Board of Admiralty had offered a prize of £20,000 for anyone who could find a passage across the northern part of Canada; in 1818 Ross had led an unsuccessful expedition to find this Northwest Passage. He made a second attempt in 1829 and was sponsored by Felix Booth. A second-hand vessel was purchased in Liverpool and grandly named *Victory* (Ross had served on HMS *Victory* during the Napoleonic Wars), towed to London and fitted up with steam engines and paddle wheels, which would be lifted from the water so that they were not damaged by ice. The high-pressure boiler and engine was designed by Ericsson. Ross and his expedition sailed from London in May 1829; this second arctic expedition was also a failure, but Ross was the first European to reach the magnetic North Pole, on 1 June 1830. The ship became trapped in ice and was abandoned, and the crew dragged the ship's boats (carrying their supplies) over the ice, eventually being rescued in August 1833 by a passing ship, the *Isabella*, which Ross had commanded back in 1819. Four members of the crew died during the expedition. Ross was knighted in 1834 and received many European decorations, including being made Knight of the Swedish Royal Order of the Polar Star. Having been seconded to Sweden during the Napoleonic Wars, Ross was British Consul at Stockholm in

1839–46. But this was no quiet retirement: in 1850, at the age of seventy-two, he undertook a third Arctic voyage, this time to search for the remains of the tragic Franklin expedition, but despite the funding and optimism of Lady Franklin, Sir John and his crew were reported dead.

Locomotives

By the time of the Rainhill Trials, the two Johns thus had considerable experience of building steam engines, and had developed a very efficient steam plant. It was therefore natural that on hearing of the Rainhill Trials they, together with Charles Vignoles, would enter their own competitor: the *Novelty*. Vignoles, while a fierce rival of George Stephenson on the Liverpool & Manchester Railway, shared George's opinion of the superiority of the locomotive. Vignoles was apparently lacking in technical knowledge of the steam engine,

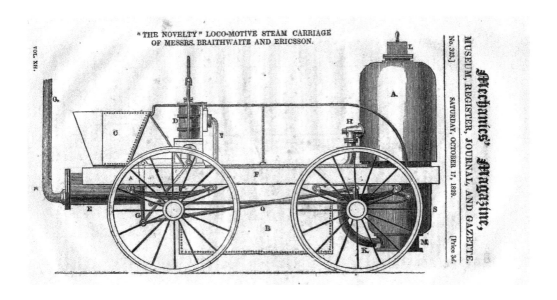

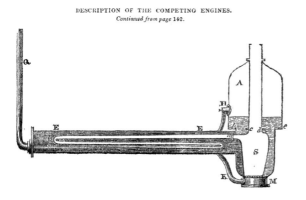

Above: *Novelty* as drawn by C. B. Vignoles for the readers of the *London Mechanics' Magazine*. The *Mechanic's Magazine* was a vocal champion of *Novelty* and her builders.

Left: *Novelty* had a theoretically brilliant boiler design, with an upright section containing the firebox and a horizontal section containing a tapering flue tube, requiring a blast of pressurised air from bellows to help the fire burn.

but 'he was a quick learner and always ready to take up a challenge'. Using his own finances, he backed Braithwaite & Ericsson and their Rainhill entry.

Given that Braithwaite and Ericsson had only seven weeks in which to build *Novelty*, and the very close similarities between the locomotive and fire pump of the same name, it is possible that the one was used to build the other. Sadly, despite all the encouragement of the judges and Vignoles' biased reportage in the *Mechanics' Magazine*,

A full-sized working replica of *Novelty* was built in 1979 for the 'Rocket 150' event, seen here after completion and undergoing testing. (Ian S Carr/Bowes Railway)

Although now based in Sweden, the replica of *Novelty* returned to the UK to take part in a re-enactment of the Rainhill Trials in 2002 and the 'Riot of Steam' event in Manchester to mark the 175th anniversary of the Liverpool & Manchester Railway. (David Boydell)

Left: In 1929 the Science Museum built a full-sized replica of *Novelty* incorporating the four original wheels and one of the original cylinders and valve gear components. It is currently on display at the Science & Industry Museum, Manchester.

Below: Braithwaite & Ericsson's next attempts were the doomed *William IV* and *Queen Adelaide*. Whereas *Novelty* used bellows to provide air for the fire, they used large fans on top of the boiler. Both were expensive failures.

William the Fourth, Locomotive Engine.

Novelty was not a success. While Braithwaite and Ericsson's steam plant had been suited to driving stationary fire pumps, applying the same technology to a moving vehicle that not only had to move itself but also carry a useful load was rather more complicated. In order to increase steam production with the locomotive *Novelty*, a pair of double-acting bellows had been used, but these absorbed most of the power output of the cylinders. Thus, *Novelty* was rebuilt with a third cylinder specifically to work the bellows and, under test in early 1830, was found to work admirably. Thus, flush with success, Braithwaite and Ericsson proposed building two even larger *Novelty*-type locomotives, but this time with the fire livened up not with bellows but by a fan driven by exhaust steam from the cylinders. Costing £1,000 each, *William IV* and *Queen Adelaide* were not a success and, despite the promises of Braithwaite and Ericsson, in fact delayed the running of the goods service by the Liverpool & Manchester Railway until early in 1831 when two Stephenson locomotives had been delivered. Ericsson found himself part of the London establishment and his work backed, perhaps uncritically, by the likes of the *Mechanics' Magazine*. His locomotive designs (*Novelty, William IV*) were praised as the only serious rival to those of Robert Stephenson; *Novelty* was the 'people's choice' because Braithwaite and Ericsson represented the 'right sort' of people. Even when *William IV* was shown not to work, the *Mechanics' Magazine* stood by Ericsson and did not blame him for any failure of the design, casting aspersions on the Liverpool & Manchester Railway instead.

Hot-air Locomotive

The hot-air engine was Ericsson's first love. He had designed his first hot-air engine in 1823 and patented it in 1826. A second design appeared in 1828, but this was not a success. Another design of 1830 was a rotary engine, which could also be worked by steam, hot air or water. A patent was obtained in 1833 for another hot-air engine, and indeed a locomotive powered by a hot-air engine was designed and the rights to build one offered to the Liverpool & Manchester Railway, but they declined.

ERICSSON'S CALORIC ENGINE.

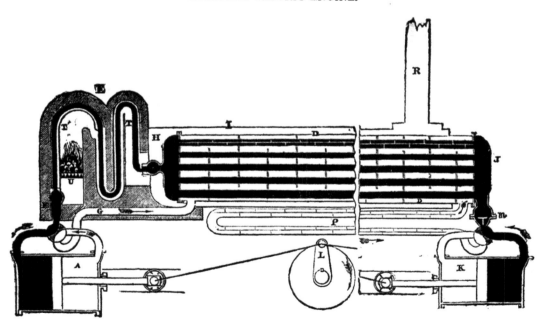

Ericsson's final locomotive design was for a 'hot air' locomotive, which used the properties of hot air (expanding) and cold air (contracting) to move a piston in a cylinder, which in turn drove the wheels.

Braithwaite, Railway Engineer

In 1834, perhaps disappointed at the repeated setbacks of his fire pumps and locomotives, John gave up running the family business and his younger brother, Frederick, took the reins. John then became a consulting civil engineer. His first project was joint with Vignoles, the surveying of the Eastern Counties Railway from London to Norwich and Yarmouth via Colchester and Ipswich in 1834. The enabling Act was passed in 1836 and Braithwaite was appointed chief engineer. The line was laid to a gauge of 5 feet, but sufficient land purchased and the earthworks, etc., built to accommodate Brunel's 7-foot-¼-inch gauge. The Eastern Counties opened from a temporary terminus in June 1839, with the new station at Bishopsgate coming into use during the following year. It was relaid to standard gauge in 1844 and became part of the Great Eastern in 1862. While engineer to the Eastern Counties he introduced many modern labour-saving machines, such as an American steam navvy and steam piledriver.

Braithwaite was engineer for the Direct Exeter Railway and was also consulted as an engineer for railway projects in France (1840–46), including the railway from Rouen to Dieppe and the French 'Western Railway' (Chemins de fer de l'Ouest). He was joint founder of the *Railway Times* (1837), which he began in conjunction with J. C. Robertson as editor and continued as sole proprietor until 1845 when he retired from railway business.

Meanwhile, Frederick Braithwaite formed a partnership with John Milner as Braithwaite, Milner & Co. in 1836 as engineers and engine builders. That firm built at least fourteen locomotives of the Bury type with bar frames and D-shaped fireboxes for the Eastern Counties Railway and a similar number for the United States. One of their locomotives, the *Rocket*, built for the Philadelphia & Reading Railroad in 1838, is on display at the Franklin Institute. After the failure of the firm in around 1845 Frederick concentrated on wells and water supply, working in Lisbon in 1848 to established a clean, modern water supply for that city, and in 1850 gave evidence to the Board of Health about the water supply in London. Like his brother he was a member of the Institute of Civil Engineers. He died aged sixty-eight in 1865.

John Braithwaite was elected a member of the Society of the Arts in 1819 and a member of the Institute of Civil Engineers in 1838. He died quite suddenly in September 1870 and is buried at Kensal Green Cemetery. His obituary notes:

[He] was always kind and hospitable; his apprentices and employees were noticed by him and liberally treated. His conversation was lively, frequently instructive, and a vein of humour appeared in his remarks.

He had no mean skill in painting and drawing, and his professional sketches were clear and explanatory. He was correct in his calculations, strict in his estimates, and his works on the Eastern Counties railway were characterised by solidity of construction.

New Horizons: Ericsson in America

Sticking with ship design, beginning in 1835 Ericsson started work on designs of screw propellers for steamships. Working simultaneously was Francis Petit Smith (1808–74), who was the first to take out a patent for a screw propeller in May 1835. Smith built an experimental vessel, the *Francis Smith*, in which to test his ideas in 1836. Meanwhile,

Ericsson built his own steam boat, the *Francis B Ogden*, a year later, which he demonstrated on the Thames in London to senior members of the Board of Admiralty. Despite achieving a speed of 10 mph the men from the Royal Navy were, sadly, unimpressed.

Although the Royal Navy had been reluctant to adopt the screw propeller, a wealthy and influential American naval officer, Commodore Robert Stockton (1795–1866), was impressed with the design and invited Ericsson to try his luck in the United States. Stockton came along at the right time as Ericsson was then in penury, and no doubt the offer of technical and financial support would have been welcome. In 1839 Ericsson packed his bags and emigrated to New York. America, and the opportunities it offered, was good to Ericsson at first. In New York, Stockton invited Ericsson to design a new type of warship, which would later be christened the USS *Princeton*. She was the US navy's first metal-hulled, screw-driven steam warship. She was laid down in October 1842 and launched in September 1843. She won a speed trial against the paddle wheeler SS *Great Western*, then believed to be the world's fastest ship.

Unfortunately, during a visit by President John Tyler in February 1844, one of the newly designed guns burst, killing Secretary of State Abel P. Upshur and Secretary of the Navy Thomas Gilmer as well as six others. During the ensuing crisis, Stockton placed the blame on Ericsson and, using his influence, was able to block the US navy from paying Ericsson for the ship. Ercisson's name was thus 'anathema during much of the remainder of the antebellum period'.

Alongside his shipbuilding, Ericsson continued with his experiments with hot air engines, building eight of them of ever-increasing size between 1840 and 1850, the last engine costing a mammoth $7,000. A ninth engine was built in 1851 at a cost of $17,000, the two cylinders of which had a stroke of 24 inches and bore of 48 inches. During the following year the king of Sweden sent him a congratulatory message for his work on hot-air engines. It was in the same year that Ericsson began one of his most audacious schemes: a 260-foot-long ship powered by hot-air engines. The keel was laid in April 1852 and the first trip made by the *Ericsson* just nine months later in January 1853. Much was made of this new venture, with one newspaper reporting that the 'age of steam is dead'. Modifications took place and further sea trials were held in spring 1854. The US navy inspected the vessel and found that the hot-air engines took far more space than conventional steam plant, were more expensive to construct and the ship was 'too slow for commercial purposes', but the fuel consumption was markedly less than a steamship. The vessel was rebuilt as a steamship in 1858 and during the American Civil War was armed with a battery of guns for war service and was eventually converted to a sailing ship.

USS *Monitor*

Ericsson's greatest claim to fame is the design of the world's first turreted warship to see action, the USS *Monitor*. Ericsson was viewed by the US navy as something of a maverick, if not with outright distrust. Ericsson's first design for a turreted warship had been presented to Emperor Napoleon III (1808–74) of France at the outbreak of the Crimean War in 1854. Although the emperor immediately recognised the advantages of the new design, the French navy was sceptical of it and there the matter sadly ended.

So, when the US navy was looking for its own ironclad to counter the Confederate CSS *Virginia*, Ericsson's design for a 'cheese box on a raft', or 'Ericsson's folly' as *Monitor* was then dubbed, there was considerable doubt whether the ship would even be able to float. *Monitor* was one of three designs submitted, and the least conventional. Initially rejected as only suitable for inland waters (which was to be proved right when she sank under tow), *Monitor* was chosen only after direct the involvement of President Lincoln: 'It strikes me there's something in it.' Construction began in October 1861 and the 120-ton vessel was launched in January 1862 and commission in February. She was armed with two 25-inch Dahlgren muzzle-loading guns and her first captain was Lieutenant John L. Worden.

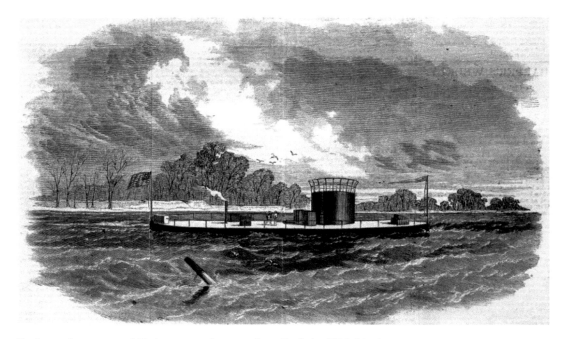

'A cheese box on a raft' is how one observer described the USS *Monitor*.

Monitor's finest hour was during the Battle of Hampton Roads. On 8 March 1862 the CSS *Virginia* attacked the blockading Union squadron, causing havoc among the wooden ships, their shot bouncing off her armoured casemate.

On the following day (9 March), *Monitor* joined the fray in what was to be the first battle of the ironclads. Neither ship could really harm each other, and both sides claimed victory.

Monitor had been designed with a very low freeboard and to operate in rivers, estuaries and coastal waters. Ericsson hadn't intended her to be seagoing and she sank while under tow. Her wreck was located in 1973.

On 6 March 1862, *Monitor* left New York with a crew of sixty-three on board and sailed to destiny to face the *Virginia* in the Battle of Hampton Roads (8–9 March 1862). The armoured *Virginia* had wrecked the US navy blockading force, but the arrival of *Monitor* turned the tide of battle in favour of the Union: it was the first battle between ironclad steam-driven warships and would prove a turning point in naval architecture. Gone were long broadsides of guns, but instead a revolving turret, capable of turning through 360°, which rendered obsolete the old 'wooden walls' and even the most up-to-date European warships such as HMS *Warrior* or *La Gloire*. Ericsson designed three more classes of *Monitor* for the US navy during the Civil War: the Passaic Class, Canonicus Class and Casco Class. But the US navy interfered with the design of the latter, resulting in a disappointing and unsuccessful warship. During the war the *Monitor* warship evolved to carry twin and quadruple turrets 'designed to be John Ericsson's ultimate expression of a true seagoing ironclad'.

After the war Ericsson designed and built Monitors for other navies and continued to design naval weapons, including an early torpedo boat that could outrun the ironclad steam warships of the day and was capable of firing a dynamite torpedo under water, yet no navy took an interest in it. He also experimented with solar power. He died on 8 March 1889 and, by his own request, was buried in Sweden. Ericsson was highly intelligent, did not suffer fools gladly, and was proud and stubborn. He was often 'the most intelligent person in the room', or at least thought he was, and with others who did not share his vision (or understand it) he could become quite volatile and form lifelong grudges. He was often working at the cutting edge of technology, which often let him down rather than his ideas. He was also let down by his finances and was frequently found insolvent and was even thrown into debtors' gaol while in London. Sadly, he was often let down by his backers and erstwhile friends. His designs, while theoretically brilliant, were often unnecessarily overcomplicated and not always a conspicuous success.

CHAPTER 6

Timothy Hackworth

Timothy Hackworth was born in Wylam in December 1786. He was the son of John Hackworth, the foreman smith at Wylam Colliery. Whereas George Stephenson apparently received only the most rudimentary of educations, the young Timothy was educated at the village school and was then apprenticed to his father to learn the trade of a smith. John Hackworth had attained local 'celebrity as a boiler builder, blacksmith and general worker in metals' and was to hold the position of foreman smith for over twenty years, until his death in 1804. It was quite natural that, according to his son John Wesley Hackworth, Timothy:

> Gave early indication of a natural bent and aptitude of mind for mechanical construction and research, and it formed a pleasurable theme of contemplation for the father to mark the studious application of his son to obtain the mastery of mechanical principles, and observe the energy and passionate ardour with which he grasped at a thorough knowledge of his art. (Young, p. 43.)

With the death of his father when just seventeen, Timothy found himself head of the household having to care for his mother, brother and five sisters. Having completed his articles in 1807, Timothy stepped into his father's position as foreman smith at Wylam.

Timothy Hackworth, the 'Superintendent of the Permanent and Locomotive Engines' on the Stockton & Darlington Railway, and pioneer locomotive builder.

Soho House, at New Shildon, which was home to the Hackworh family. Today it forms part of the Locomotion Museum.

Wylam Colliery was owned by Christopher Blackett, who was also lord of the manor. The day-to-day management was vested in the viewer William Hedley, but Blackett also had a mechanical turn of mind. Coals from the colliery were taken down to the Tyne on a wooden waggonway. Laid to a gauge of 5 feet, it had opened in 1748, making it one of the older lines on Tyneside. It was relaid with iron plate rails in 1808 and during the following year Blackett tried to persuade Richard Trevithick to build him a locomotive, but just when a new patron might have been found, Trevithick declined to do so. Importantly, however, Blackett was the 'only person to show a practical interest in the locomotive between Trevithick's ... *Catch Me Who Can* of 1808, and Blenkinsop's ... engine of 1812'. A locomotive on Trevithick's principles was later built and dubbed *Black Billy*. It was a single-cylinder machine with a flywheel and geared drive constructed in 1813 by Thomas Waters of Gateshead, who was the local agent for Trevithick's patent high-pressure steam engines. According to Nicholas Wood, writing in 1825, 'For some time ... the whole of the coals was taken down the Rail road by this locomotive.' Wylam Colliery was never particularly rich and, like other coal owners, Blackett was keen to lower his working expenses, especially so due to the rising prices of not just horses but their feed, and also the men's wages. *Black Billy* had shown the way and three further locomotives were built at Wylam: *Elizabeth*, *Jane* and *Lady Mary*, the first two being better known to history as *Puffing Billy* (so-called because of William Hedley's asthma) and *Wylam Dilly*. These locomotives began work in 1814; Christopher Blackett notes in March of that year that the first engine was, or would soon be, at work on a trial basis. The actual date of introduction of steam power on the Wylam Waggonway is shown by the dramatic cost in the increase of repairs to the track in August 1814, rising from £5 per fortnight to as much as £47 per fortnight in January 1815 before reducing to £10 per fortnight in April 1815. As Andy Guy has argued this sudden increase in the cost of maintaining the permanent way was most likely due to track damage by the Wylam engines, suggesting that one or both began work in August 1814 – only a month after Stephenson's *Blucher* (July 1814). The final payments for the locomotives were made in 1816.

Although the role of Hedley and Hackworth in the design and construction of these three locomotives has been often overemphasised by their families, who was ultimately responsible for the design is unclear. Hackworth's grandson, Robert Young, claims that Hackworth was the instigator of *Black Billy* and almost single-handedly designed *Puffing Billy* and her two sisters, while Jonathan Foster reported to Samuel Smiles he alone was responsible. Certainly, Christopher Blackett had a mechanical turn of mind and so too Hedley, with both having an interest in steam traction. Given the similarities between the Wylam engines and Chapman & Buddle's eight-wheeled bogie locomotives for Lambton Colliery of 1814, this suggests that they too influenced the design of the Wylam locomotives. It's also interesting to note that Hedley was in dispute with Chapman and Buddle at this crucial period. Furthermore, so happy was Hedley with the design that he still recommended the design in the 1830s.

The working replica of *Puffing Billy* on the Pockerley Waggonway at Beamish Museum gives an excellent impression of the first steam railways during the age of Jane Austen. (Rob Langham)

Methodist Preacher

While all this technical innovation was going on at Wylam, Timothy left the village of his birth under a black cloud. Around 1812 he had started attending services held by the Methodists, a religious movement founded by Revd John Wesley. While the Quakers, such as the influential Peases of Darlington or the Unitarians like Henry Booth, were still discriminated against in law for not being members of the Church of England, they had, despite their Nonconformity, attained a certain semblance of respectability, particularly as bankers, merchants and engineers. Many had become quite wealthy in the process. Methodists, then known as 'ranters', were considered to be 'beyond the pale' and were not yet 'respectable'. In 1814 Hackworth married Jane Golightly, who sadly had to leave her family home because of her religious beliefs – her parents being staunch Anglicans. Timothy named his eldest son John Wesley Hackworth (1820–91) in honour of the founder of his faith. Then, in *c.* 1815, Hackworth had to leave Wylam because of his Methodist activities. He was a lay preacher and 'missioner'; he refused to work on Sundays and got into trouble for handing out religious tracts to his fellow workmen. Happily, he received an offer of employment as foreman smith at Walbottle Colliery, where he moved in 1816. Despite his religious zeal, Hackworth was no sectarian and, rather like George Stephenson, was determined to give his children the best start in life possible. From an advertisement in the *Newcastle Journal* (25 August 1838) 'Mr Timothy Hackworth, Engineer, Shildon' is mentioned as having sent his sons to the Classical and Commercial School in Darlington, run by the Anglican minister Revd Joshua Wood MA, the fees of which were between 30 and 40 guineas per year. A daughter was educated in Belgium and married a French Catholic.

Hackworth's time at Walbottle was apparently unremarkable, but both there and at Wylam he gained considerable experience with colliery pumping and winding engines.

The former Soho Wesleyan Chapel of 1865, Shildon, replaced an earlier structure built in 1830. The Hackworths were great supporters of Wesleyanism.

In early 1824 Robert Stephenson & Co. were looking for experienced engineers to take on the technical direction of the firm, and Hackworth was recruited – probably as a foreman. While working at Forth Street, he possibly designed an outside cylinder locomotive reminiscent of those at Wylam. Hackworth left Robert Stephenson & Co. at the end of the year and, thanks to the recommendation of George Stephenson, was appointed as the Superintendent of the Permanent and Locomotive Engines on the Stockton & Darlington Railway from May 1825 with an annual salary of £150 and 'the company to find a house, and pay for his house, rent and fire'. It was his job to keep the locomotives and fixed plant in working order, and eventually he became the running superintendent, keeping the trains moving too. It shows considerable trust and the tolerance that Quakers are known for that the S&D appointed Hackworth – someone considered an outsider – to such an important position. But then Quakers, too, were also 'outsiders'. During 1829 he entered designs for coal staithes at Middlesborough for the S&D Railway, for which he was awarded a prize of 150 guineas in April of that year; construction began in September. Having already successfully competed in one engineering competition no doubt buoyed Hackworth's confidence for entering the Rainhill Trials. Perhaps some of his prize went toward the cost of building *Sans Pareil*. Imagine then his disappointment when, coming on the back of failure at Rainhill, in November the Tees Navigation Company issued a cease-and-desist order to the S&D, which curtailed work on the new coal drops. This was only able to resume in the following year after the passage of an Act of Parliament. The coal drops came into use on 27 December 1830.

The coal drops on the River Tees at Middlesborough, then called Port Darlington, were designed by Timothy Hackworth in 1829.

Locomotive Builder

The poor build quality and poor steaming of the earliest Stockton & Darlington locomotives resulted in Hackworth seeking permission to build his own: the *Royal George*, which first steamed in September 1827. To do so he recycled many of the parts of an earlier machine built by Robert Wilson (1781–?) of Gateshead named *Chittaprat*. Wilson's locomotive had been the first to use four cylinders – two on each side of the boiler yoked together working a common connecting rod to the driving wheels. It ran on two-part, cast-iron wheels called 'plug wheels'. The inner part of the wheel was turned true on a lathe, and the outer, which was too large to be turned on the lathes of the day, was secured using wooden plugs into which iron pins were driven, expanding the plugs and securing both halves of the wheel. Although Robert Young claims Hackworth was the progenitor of these wheels, their design is due to Robert Wilson for his locomotive *Chittaprat*. Similarly, while Hackworth adopted a sprung safety valve for the boiler of *Royal George* using a stack of leaf springs, this was not a new idea as Blenkinsop & Murray in 1812 used a sprung safety valve.

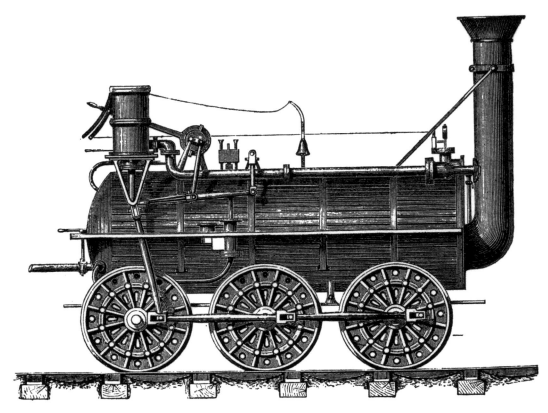

Hackworth's first locomotive was the *Royal George*, an impressive 0-6-0, which utilised the boiler from the ill-fated *Chittaprat*. It was possibly the most powerful locomotive in Britain when it was built.

Royal George was carried on six wheels and had a large boiler, which, thanks to a return flue, provided a larger heating surface than the contemporary Stephenson locomotive. The wheels were coupled with outside connecting rods – an innovation of James Kennedy. In a marked advance from the Stephenson locomotives with vertical cylinders part-immersed on the centre line of the boiler and complicated Freemantle parallel motion, *Royal George* had cylinders either side of the boiler. They were more compact so made better use of the steam and were inverted, so that they drove downward and acted on crank pins on the rearmost set of wheels. This was a major breakthrough, but vertical cylinders driving downwards increased the hammer blow effect on the track and also meant that the driven pair of wheels could not be fitted with springs. While Robert Stephenson overcame this by using inclined cylinders, and later horizontal cylinders under the smokebox, Hackworth stuck to vertical, inverted cylinders for most of his career – the old belief that the weight of a piston in a horizontal cylinder would wear the bore oval died hard. *Royal George* was ordered to be put on springs in October 1828, the springs acting on the leading and middle axle. Hackworth's second locomotive, *Victory* of 1829, was a more refined version of *Royal George* and design features of both were incorporated in *Sans Pareil* for the Rainhill Trials. In building *Sans Pareil*, Hackworth had invested all his working capital and he cautiously wrote to the board of the Liverpool & Manchester Railway seeking reassurances in case his entry 'should be very nearly as complete & good in all respects as to the one which should win the premium' and whether they would be prepared to purchase the engine. Henry Booth replied that the board would 'deal liberally' with the 'proprietor of an Engine under those circumstances' but were not under any obligation to purchase it. Ultimately the board did consent to purchase *Sans Pareil* for £550, and she was put to work on the Bolton & Leigh Railway because she 'will not work on Coke, and therefore unfit for the Liverpool & Manchester Line'.

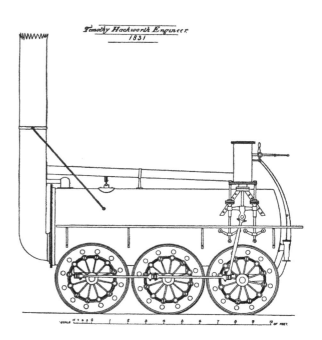

Victory was a development from *Royal George*, Note the use of a cross head to guide the piston rod rather than parallel motion. The valves were worked by eccentrics on the driven axle.

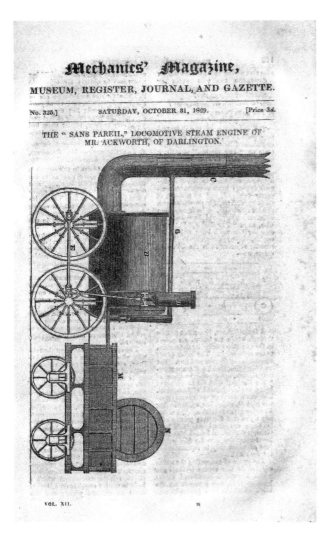

Mechanics' Magazine,

MUSEUM, REGISTER, JOURNAL, AND GAZETTE.

No. 325.]   SATURDAY, OCTOBER 31, 1829.   [Price 3d.

THE " SANS PAREIL," LOCOMOTIVE STEAM ENGINE OF
MR. ACKWORTH, OF DARLINGTON.

VOL. XII.   M

Hackworth's Rainhill entry, *Sans Pareil* (meaning without equal), as drawn by the *Mechanics' Magazine*. Unfortunately, the tender was drawn at the wrong end!

There had been various claims as to who invented the blast pipe, and indeed Hackworth's son (J. W. Hackworth) and grandson (Robert Young) claim that it was Timothy. This is despite the fact Richard Trevithick used and described the effect of such a device in 1804. George Stephenson had adopted it in 1814 and noted the same effect as Trevithick had. Goldsworthy Gurney, the Quaker scientist, engineer and inventor, also claimed to have originated the blast pipe (1822). J. W. Hackworth went as far to say that Stephenson's '*père et fil*' never used the blast pipe, but this is untrue. He also suggests that at Rainhill some of Stephenson's men sneaked into the Hackworth camp and dismantled *Sans Pareil*. They copied the blast pipe, had a pattern made, rushed the pattern to a foundry over 10 miles away near Wigan and had one cast, then fitted the blast pipe to *Rocket,* which was then at John Melling's engineering works at Rainhill, despite Melling's works not existing until 1840. This was all accomplished in one night! It's an outlandish claim with no basis in fact, and one in which Hackworth was nearly taken to court for libel.

The original *Sans Pareil* is nowadays on display at Locomotion Museum, Shildon. Its present appearance is thanks to an extensive Victorian restoration.

For 'Rocket 150' in May 1980 a working replica of *Sans Pareil* was built by British Railway Engineering Limited Apprentices. She is seen here at Bold Colliery.

Sans Pareil rattles across the historic Water Street Bridge at the Science & Industry Museum, Manchester, as part of the 2005 'Riot of Steam'. (David Boydell)

Sans Pareil raising steam at Manchester. (David Boydell)

Different types of boiler had different types of blast requirements. Stephenson's straight-flue boilers didn't need a strong blast requirement; in fact, it would be detrimental, but in order for them to make as much steam as the more efficient return flue it would have needed a more pronounced 'blast'. Such a boiler, however, would have required a sharper blast characteristic than a multi-tubular boiler to make the same amount of steam. Thus, *Sans Pareil* would have had a far more pronounced 'chuff' than *Rocket*. At Rainhill, of course, the emphasis was on prevention of anything that might cause a 'nuisance' to the anti-railway lobby – hence the use of coke to prevent smoke. And the relatively soft blast of *Rocket* would promote fuel efficiency and reduce noise.

Soho Works

Hackworth was largely responsible for developing the first engineering works for the Stockton & Darlington Railway (1825–30). Having had to overcome several practical issues of working a railway, along with technical and mechanical issues, by 1830 he had developed a type of locomotive that was reliable, robust and capable of doing the task asked of it in its local niche. Despite the majority of his designs being heavy, six-wheeled, vertical cylinder mineral engines, he was not afraid to experiment; for example, his 1830 design for an inside, horizontal-cylinder passenger engine, the *Globe*. By 1832 his salary had been increased from £150 to £200 per annum and during 1833 to 1834 the S&D decided to contract for not only engineering services but the running of its trains too. Hackworth became one of the principal contractors on the S&D, putting Timothy in the position of being both a contractor and employee, while at the same time developing his own independent engineering business.

Remains of Hackworth's Soho Works at Shildon, now part of Locomotion Museum.

In order to finance his own railway workshops, he took out a substantial loan of £9,000 from the Quaker proprietors of the S&D. A company was established known as Hackworth & Downing and the new Soho Works were managed by Timothy's bother Thomas. The partnership was dissolved in 1837 and the company continued to trade as Thomas Hackworth & Co. until 1840. It's not clear how good a businessman Hackworth was as, in addition to his £9,000 start-up loan, he took out a further loan with the Quaker bank Johnathan Backhouse & Co. around July 1837.

Hackworth also built one of six locomotives made for the first public railway in Russia in 1836: the Tsarskoye Selo Railway, which opened on 30 October 1837. A sixteen-year-old John Wesley Hackworth was sent to Russia to help reassemble and fettle the engine once there. Often described as the first locomotive to run in Russia, that honour in fact goes to a Russian-built locomotive, the *Cherepanov* of 1834, which copied Robert Stephenson's Planet design. Hackworth's locomotive was probably built under subcontract from Robert Stephenson & Co. Stephenson's did not have the capacity to fulfil the entire order and only two locomotives were built by them; one was built by Cockerill of Belgium and two by Charles Tayleur at Newton-le-Willows. They were all built to Stephenson's 2-2-2 Patentee design to 6-foot gauge. According to the diary left by one of Stephenson's men, Thomas Wardropper, Hackworth's engine was first tried on 15 November 1836 but 'she made a very bad start from the hurried manner she was put together, after she had run about 60 or 80 yards she was put by...' Of John Wesley Hackworth, Wardropper commented that

Timothy Hackworth's business card, depicting his unique 0-4-0 locomotive *Globe*, which had cylinders mounted beneath the firebox driving a crank axle, and a spherical steam dome. (After Young, 1923)

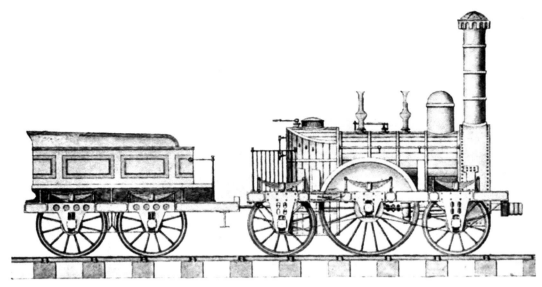

HACKWORTH'S LOCOMOTIVE FOR RUSSIA, 1836.
(From an old drawing)

Hackworth built a locomotive under sub-contract from Robert Stephenson & Co. for the Tsarskoye Selo Railway, which opened in 1837. (After Young, 1923)

he 'hardly knows how many wheels they [the engines] have got'. The first Stephenson engine was completed by December and made 'a very fine start and ran at a rate of 30 miles per hour' while Cockerill's locomotive 'made a very stiff start, got steam badly'. The weather became cold and snowy, causing many problems for the British locomotive builders and when one of Hackworth's cylinders cracked from the cold 'they took it out and I [Wardropper] took it to [St] Petersburg to get it mended'. The three engines built by Hackworth for the Albion Coal Mining Company of Nova Scotia in 1839 were amongst the earliest but not the first steam locomotives to run in Canada. The first locomotive in Canada was the *Dorchester* built by Stephenson in 1836.

During 1839 the S&D management decided to terminate their agreement with Hackworth for their New Shildon workshops and in April of that year Hackworth gave a year's notice. Thus, in May 1840, Hackworth moved his own operations to Soho Works, trading as Soho Engine & Foundry Works, repairing locomotives of the S&D as well as other lines. Thomas Hackworth established his own business with George Fossick, who provided much of the capital to start this new partnership based in Stockton. By the mid-1840s the Soho Works was in decline, thanks in part to a national downturn in the economy, which badly affected the locomotive building industry. The works were also old fashioned with great reliance on hand, rather than machine, tools. This problem was made worse by Hackworth overstretching himself with a large contract for the London, Brighton & South Coast Railway. Getting into financial difficulties, Hackworth found himself at the Durham Assizes for debt.

His final locomotive design, a 2-2-2 named *Sanspareil No. 2*, was completed in 1849, a version of David Joy's 'Jenny Lind' design. It had 6-foot 6-inch driving wheels and a welded boiler, built to a joint patent taken out by Timothy and John Wesley Hackworth.

Hackworth's locomotive designs, although seemingly old fashioned with return-flue boilers and vertical cylinders, were cheap to build, cheap to operate, and ideally suited to working heavy coal trains.

It was apparently built as a speculative venture, but there is a sense of desperation in the challenge issued by J. W. Hackworth to Robert Stephenson:

> It is now about 20 years since the competition for the premium of locomotive superiority was played off at Rainhill, on the Liverpool and Manchester Railway. Your Father and mine were the principal competitors. Since that period you have generally been looked to by the public as standing first in the construction of locomotive engines. Understanding that you have now running on the York, Newcastle and Berwick Railway a locomotive engine which is said to be the best production that ever issued from Forth Street Works, I come forward and tell you publicly that I am prepared to contest with you, and prove to whom the superiority in the construction and manufacture now belongs. At the present crisis, when any reduction in the expense of working the locomotive engine may justly be hailed as a boon to railway companies, this experiment will no doubt be regarded by them with deep interest, as tending to their mutual advantage. I fully believe that the York, Newcastle and Berwick Railway Company will willingly afford every facility towards the carrying out of this experiment.

Robert Stephenson wisely declined to respond. Timothy Hackworth died in 1850, apparently with significant debts, and his wife Jane followed him to the grave two years later. John Wesley Hackworth commenced his own engineering works in Darlington and the Soho

Works were put up for sale in 1852, finally being sold to the S&D in 1855. The site later became part of Shildon Gas Works and is today part of the NRM outstation at Locomotion.

Timothy Hackworth was a skilled, intuitive engineer. It would be unfair to Hackworth to see him as a failure because of Rainhill or to compare him with Robert Stephenson because of their Rainhill rivalry. Hackworth was a competent engineer, but, other than two locomotives in Canada and one in Russia, didn't work on the national or international stage. He was an important player in the evolution of the locomotive during the late 1820s and early 1830s; thereafter, his designs stagnated somewhat, stubbornly sticking to vertical or angled cylinders and return flue boilers. That said, however, Hackworth's locomotives were cheap, simple to build, cheap to run and excelled at the task for which they were designed: pulling very heavy coal trains at low speeds on often steeply graded track. And, in the niche where they evolved, they were undoubtedly a success.

Hackworth's final locomotive design was *Sans Pareil II*, largely based on the contemporary Jenny Lind design of David Joy, completed in 1849. (After Young, 1923)

CHAPTER 7

They Also Ran

Thomas Shaw Brandreth

Brandreth (1788–1873) is best known as a classical scholar and solicitor, with a legal practice in Liverpool. He was elected a Fellow of the Royal Society in 1821 for his work on mathematics. He was of an inventive turn of mind and, like Ross Winans (see p. 91), also designed and patented a form of 'anti-friction wheel' in 1825. Examples of Brandreth's

Literal horse power! Thomas Shaw Brandreth's entry for Rainhill was the 2-horsepower *Cycloped*. The second horse is not pictured.

patent waggons were tested by the Liverpool & Manchester Railway, where they did not find favour, but were used by the Stockton & Darlington Railway and the Bolton & Leigh Railway.

Brandreth is also credited with the invention of the 'Dandy Car[t]' used on horse-drawn railways. Once the horse had dragged their load to the top of a hill, they were trained to ride down-hill in the 'Dandy Car', thus affording them a period of rest and recuperation. It's no surprise, therefore, that Brandreth's Rainhill entry, the *Cycloped*, was literally a 2-horsepower machine, powered by two horses walking on a treadmill, which, via gearing, drove the wheels. He patented this 'method of applying animal power to machinery' on 9 September 1829, a month before it made its debut at Rainhill. Brandreth was also a director of the Liverpool & Manchester Railway, but resigned in 1830 before the line opened. He retired to London to work on translations of Homer. He died in Worthing in 1873.

Timothy Burstall

Burstall was born in Lincolnshire in 1776, but by 1806 had moved to London. His interest in steam locomotives stemmed from several attempts to build a workable steam coach. Together with John Hill of Leith he built a steam coach, for which a patent was granted on 3 February 1824. Steam coaches were all the rage. The first steam-propelled road vehicle had been built by Major Nicholas Cugnot in France in 1779, and Richard Trevithick had built a passenger-carrying steam coach in 1803. Several engineers were interested in the problem of road locomotion, including Goldsworthy Gurney. His first design consisted of a four-wheel steam carriage designed to pull an ordinary road coach. His second, and more successful, design was a six-wheeler. It was the first steam coach to carry fare-paying passengers. The design had a vertical boiler mounted on a four-wheel power bogey, which articulated with the coach body. The third pair of wheels at the front provided the steering. Before the growth of early mainline railways, the steam carriage was considered the up-and-coming technology of the age.

The design of Burstall's Rainhill entry, *Perseverance*, was based upon his second patent steam carriage. Overall, it resembled the steam motor part of his road coach. *Perseverance* was a four-wheeler with a vertical boiler and two cylinders. It was nearly disqualified from the Trials as, contrary to the trial rules, she was not fully sprung and nor was she fitted with a mercurial pressure gauge. The locomotive was damaged en route and Burstall spent all week repairing his machine. When she finally did run her performance was far below that required by the Trials. The poor showing of *Perseverance* at Rainhill didn't stop Burstall from experimenting with locomotives. In January 1830 he wrote to the board informing them he was building an improved locomotive. He wrote again in May 1830, but it's not clear if it was ever finished, or indeed ran. Thereafter, Burstall continued as a civil engineer and later as a 'Dealer of Patents'. He died in Glasgow in 1860 aged eighty-four. *Perseverance* could be described as the ancestor of a long line of vertical-boiler, four-wheel shunting engines as typified by the products of De Winton in Caernarfon.

Timothy Burstall's entry at Rainhill was the ill-fated *Perseverence*. Although not a conspicuous success, she could be considered the ancestor of vertical boiler locomotives.

Bury and Kennedy

Edward Bury (1794–1858) and James Kennedy's (1797–1886) joint entry, the *Dreadnought*, was not completed in time to take part at Rainhill, but it was subsequently hired to the Liverpool & Manchester Railway on ballast duties and later worked on the Bolton & Leigh Railway. It was a 0-6-0 with a return-flue boiler and inclined cylinders working a crank shaft. Mounted on the crank shaft was an accelerating wheel, and final drive to the wheels was via chains. Bury was born in Salford in 1794 and educated at the Blue Coat School in Chester. By 1825 he was the joint proprietor of a steam-powered saw mill at Toxteth, Liverpool. Kennedy was a Scot who had trained as a millwright and later as a marine engineer. For some eighteen months he had been employed by Robert Stephenson & Co. in a supervisory position, where he gained considerable knowledge of locomotive building. This would stand him and Bury in good stead when they commenced building their own locomotives at the Clarence Foundry in Liverpool and came to rival Stephenson during the 1830s in the locomotive-building field. A bitter Robert Stephenson noted in 1833 that he regretted Kennedy had ever been taken on by him at all – without Kennedy's time at Forth Street 'Bury would never have made an engine'. Bury and Kennedy's second locomotive was the aptly named *Liverpool*. Kennedy, together with Timothy Abraham Curtis and John Vernon, were taken into the partnership in 1842 as Bury, Curtis and Kennedy.

The Clarence Foundry was perhaps one of the biggest and best-equipped engineering workshops in Britain. In the late 1840s it was described as being able to turn out one locomotive and tender every week. Bury specialised in small, four-wheel locomotives being delivered as 2-2-0s for passenger work and 0-4-0s for goods trains and maintained that six wheels were unnecessary into the late 1840s. *Coppernob*, at the National Railway Museum

Bury and Kennedy's second locomotive was the aptly named *Liverpool*. Seen here in her second rebuilt state, she was the progenitor of the Bury type locomotive, which found much favour in the United States.

in York, is a surviving example of Bury's locomotive practice. In total the firm produced 415 locomotives. Thanks to the activities of an anti-monopoly, and indeed anti-Stephenson, group of directors of the London & Birmingham Railway, led by the Cropper family of Liverpool, Edward Bury enjoyed the unique position of being Locomotive Superintendent, locomotive contractor and a private locomotive manufacturer. He resigned in 1846 to become Locomotive Superintendent of the Great Northern Railway, from which position he was dismissed in 1848 for fraudulently offering contracts to his own firm – at higher prices than other contractors.

The Clarence Foundry also built marine engines and had its own shipyard where it built and repaired iron-hulled ships. Among its marine products were steam packet boats for the Dublin–Holyhead service and a pair of iron steam frigates – one for the Russian navy and one for Prussia. With the collapse of iron shipbuilding on the Mersey in the late 1840s and sustaining heavy losses on a bridge contract in Russia, the Clarence Foundry closed in 1851. Bury died seven years later in Scarborough.

Ross Winans

Winans (1796–1877) is one of those nineteenth-century engineers who defies easy characterisation. Despite being described by his employers, the Baltimore & Ohio Railroad, in 1844 as an 'ingenious mechanic', his vertical-boiler 'Grass Hopper' locomotives were obsolete the day they were built. In 1828 he developed an early form of roller bearing or 'anti-friction wheel' designed to reduce the friction of railway axles. This was on display at Rainhill, and waggons fitted with his unique type of bearing were trialled on the Liverpool & Manchester Line, which purchased twelve of them. In a display of showmanship, Winans loaded one of his waggons with 500 lb of scrap iron and pulled it along the line with a piece of twine. A second patent was taken out in 1831. His 'anti-friction wheel' was heralded as a great leap forward in the United States and indeed one observer believed that 'there will be no use for Locomotive Engines where one horse can move so much'. Despite this praise the 'anti-friction wheel' had two major drawbacks: being more expensive and wearing very quickly compared to simpler bearings. Winans is also credited with appreciating the need for coned railway wheel treads in the United States; George Stephenson in Britain had come to a similar conclusion a few years earlier.

Winans's Rainhill entry was the man-powered *Manumotive*. It was 'worked by two men who turned a windlass which actuated the wheels'. Another press description describes how it was 'cranked by a winch and lever' and was thus able to carry six passengers, 'but at no great velocity'. The Liverpool & Manchester directors were 'taken with the idea of Winans's man-propelled carriages', and they had a report written on their 'adaptability of passenger traffic'.

Back in the United States, Winans established his own locomotive-building firm in 1835 and developed a series of unusual locomotives, including his aesthetically challenged 0-8-0 coal-burning 'Camel' locomotives where the driver's cab was mounted on the middle of the boiler; or the equally peculiar 'Mud Digger', which had a geared drive. In business he has been described as something of a bully, who attempted to patent everything he did. During the American Civil War he supported the Confederacy and was nearly shot by Union forces. For a while he served as US Vice Consul in Russia, and obtained contracts to build locomotives for the railway from St Petersburg to Moscow.

William Crawshay II

William Crawshay II (1788–1862) of the Cyfartha Ironworks in Merthyr Tydfil was of a mechanical turn of mind and had invited Goldsworthy Gurney, the builder of pioneer steam road locomotives, to bring one of his road engines to Wales to try on the plateway at the ironworks. This inspired him to build his own locomotive, as he wrote to the Liverpool & Manchester Board on 7 September 1829 – only a month before the Trials were to be held – requesting particulars of 'the Stipulation and Conditions which the Premium was offered for the most improved Locomotive Engine'. Although not built in time for Rainhill, this locomotive was finally offered to the Liverpool & Manchester at the end of March 1830. Although lost in the mists of time, this enigmatic machine was run at the Cyfartha Ironworks, where its weight damaged the brittle, cast-iron tram plates. Sadly, little else is known about it. It had probably been built by Joseph Tregelles Price of Neath Abbey Ironworks. The Neath Abbey Works had a representative at Rainhill in the person of Henry Habberley Price, one of its partners.

A second locomotive from South Wales, the *Speedwell*, was also offered to the L&M. Built by Neath Abbey Ironworks to a design of Thomas Prothero, it was based on 'Braithwaite and Ericsson's plan but with alterations and improvements'. Its only similarity to *Novelty* was the use of vertical cylinders and bell cranks for the final drive.

The town of Rainhill remains proud of its railway history and links to the world-changing events that took place there over 190 years ago.

Bibliography and Further Reading

Books

Bailey, M. R. (ed.), *Robert Stephenson – The Eminent Engineer* (London: Routledge, 2017).

Booth, H., *Henry Booth. Inventor; Partner in the Rocket; Father of Railway Management* (Ilfracombe: Arthur H. Stockwell Ltd, 1980).

Dawson, A. L., *Before Rocket: The Steam Locomotive up to 1829* (Horncastle: Gresley Books, 2020).

Dawson, A. L., *Locomotives of the Liverpool & Manchester Railway* (Barnsley: Pen & Sword Transport, 2021).

Dawson, A. L., *Locomotives of the Victorian Railway: The Early Days of Steam* (Stroud: Amberley Publishing, 2019).

Dawson, A. L., *The Rainhill Trials* (Stroud: Amberley Publishing, 2018).

Dendy Marshall, C. F., *A History of Railway Locomotives Down to the End of the Year 1831* (London: Locomotive Publishing Co., 1953).

Holt, R. V., *The Unitarian Contribution to Social Progress in England* (London: Lindsay Press, 1952).

Jacob, M. C., *The First Knowledge Economy* (Cambridge: Cambridge University Press, 2014).

John, A. H., *A Liverpool Merchant House* (London: George Allen & Unwin Ltd, 1959).

Lewis, M. J. T., *Steam on the Sirhowy Tramroad and its Neighbours* (R&CHS, 2020).

Rattenbury, G., and Lewis, M. J. T., *Merthyr Tydfil Tramraods and their Locomotives* (R&CHS, 2004).

Rolt, L. T. C., *George and Robert Stephenson* (London: Pelican Books, 1978).

Smiles, R., *Memoir of the Late Henry Booth* (London: Wyman & Sons, 1869)

Smiles, S., *The Life of George Stephenson* Centenary Edition (London: John Murray, 1881)

Thulesius, O., *The Man Who Made the Monitor* (Jefferson: McFarland & Co., 2007).

Young, R., *Timothy Hackworth and the Locomotive* (London: Locomotive Publishing Co., 1923).

Papers

Bailey, M. R., 'Blücher and After' in A. Coulls (ed.), *Early Railways 6* (Six Martlets, 2019), pp. 79–102.

Bailey, M. R., 'George Stephenson – Locomotive Advocate' in *Transactions of the Newcomen Society*, Vol. 52 (1980–81), pp. 171–207.

Bailey, M. R., 'Robert Stephenson & Co. 1823–1829' in *Transactions of the Newcomen Society*, Vol. 50 (1979–80), pp. 252–91.

Davidson, P., 'Early Locomotive Performance' in Coulls, *Early Railways 6*, pp. 124–46.

Dawson, A. L., 'Success and Failure in Early Mainline Railway Management: the Liverpool & Manchester and Leeds & Selby Railways', R&CHS Railway History Research Group, *Occasional Paper 20* (May 2020).

Guy, A., 'Early Railways: Some Curiosities and Conundrums' in M. J. T. Lewis (ed.), *Early Railways 2* (The Newcomen Society, 2003), pp. 64–78.

Guy, A., 'North-Eastern Locomotive Pioneers 1805–27: A Re-assessment' in A. Guy & J. Rees (ed.), *Early Railways* (The Newcomen Society, 2001), pp. 117–44.

Hartley, R. F., 'Why Killingworth?' in Coulls, *Early Railways 6*, pp. 25–40.

Hopkin, D., 'Timothy Hackworth and the Soho Works, *c.* 1830–1850' in G. Boyes (ed.), *Early Railways 4* (Six Martlets, 2010), pp. 280–301.

Rees, J., 'The Stephenson Standard Locomotive 1814–1825' in Lewis, *Early Railways 2*, pp. 177–201.

Sellers, I., 'William Roscoe, the Roscoe Circle, and Radical Politics in Liverpool, 1787–1807', *Journal of the Historic Society of Lancashire & Cheshire*, Vol. 120 (1968), pp. 45–62.

Acknowledgements

This book was written in the midst of the Covid-19 pandemic, which meant access to primary sources in museums, libraries and archives was very much curtailed, making one reliant upon sources and information already in print. As ever, I'd like to thank Andy Mason for being his usual tower of strength during the writing process as well as to the 'MOSI gang' (Ian Hardman, Richard Garside, Cameron McTigue and Daniel Lohrenz) for their continued encouragement and friendship. Also to Stephen Weston for editing and proofreading the manuscript; fellow early railways historian Rob Langham for his friendship, encouragement and photographs; Matthew Jackson, David Boydell, Lauren Jaye Gradwell, Robert Kitching, Paul Dawsonand and Beth Furness for photographs; and finally to the Liverpool & Manchester Railway Trust for use of images. Thanks also to Andy Mason, and Stephen Weston for proofreading this text.